Microbiology: The Study of Microorganisms

Microbiology: The Study of Microorganisms

Justin Clifford

STATES
ACADEMIC PRESS
www.statesacademicpress.com

States Academic Press,
109 South 5th Street,
Brooklyn, NY 11249, USA

Visit us on the World Wide Web at:
www.statesacademicpress.com

ISBN: 978-1-63989-354-6 (Hardback)

Cataloging-in-Publication Data

Microbiology : the study of microorganisms / Justin Clifford.
p. cm.
Includes bibliographical references and index.
ISBN 978-1-63989-354-6
1. Microbiology. 2. Microorganisms. 3. Biology. I. Clifford, Justin.
QR41.2 .M53 2022
579--dc23

TABLE OF CONTENTS

PREFACE

The purpose of this book is to help students understand the fundamental concepts of this discipline. It is designed to motivate students to learn and prosper. I am grateful for the support of my colleagues. I would also like to acknowledge the encouragement of my family.

The study of acellular, unicellular and multicellular microorganisms is known as microbiology. These microorganisms can be eukaryotic and prokaryotic. The various sub-fields of this discipline are mycology, virology, bacteriology, protozoology, phycology and parasitology. The integration of these fields leads to the formation of the field of microbiology. Microscopy, staining, identification based on DNA sequencing like the rDNA gene sequence are some of the techniques used in this field for the identification of microorganisms. This field of study is used for the treatment of cancer, digestion problems in humans, biodegrading industrial and agricultural wastes and for the production of multiple biopolymers such as polyesters and polyamides. This book elucidates the concepts and innovative models around prospective developments with respect to this field. It picks up individual branches and explains their need and contribution towards the development of microbiology. Coherent flow of topics, student friendly language and extensive use of examples make this an invaluable source of knowledge.

A foreword for all the chapters is provided below:

Chapter - Introduction

Microorganisms are microscopic organisms that are unicellular, multicellular or acellular. Microbiology is concerned with the study of the structure, classification and function of these microorganisms. This is an introductory chapter which will briefly introduce all the significant aspects of microbiology.

Chapter - Subfields of Microbiology

Microbiology is divided into a number of Subfields namely mycology, bacteriology, protozoology, virology, microbial genetics, cellular microbiology, food microbiology, industrial microbiology and soil microbiology. This chapter has been carefully written to provide an easy understanding of these different branches of microbiology.

Chapter - Classification of Microorganisms

Microorganisms are classified into various types based upon their cellular composition, morphology, mean of locomotion, and reproduction. Their types are bacteria, fungus, virus, protozoa, archaea and microalgae. This chapter has been carefully written to provide an easy understanding of these classifications of microorganisms.

Chapter - Fundamental Concepts in Microbiology

The fundamental concepts of microbiology include sterilization and disinfection, microbiological culture, microbial metabolism, microbial cooperation and microbial loop. The chapter closely examines these key concepts of microbiology to provide an extensive understanding of the subject.

Chapter - Important Theories in Microbiology

Theory of abiogenesis, theory of biogenesis, microbial theory of fermentation, hologenome concept of evolution and germ theory of disease are some of the theories that fall under the domain of microbiology. This chapter has been carefully written to provide an easy understanding of these theories of microbiology.

Justin Clifford

Introduction

CHAPTER 1

Microorganisms are microscopic organisms that are unicellular, multicellular or acellular. Microbiology is concerned with the study of the structure, classification and function of these microorganisms. This is an introductory chapter which will briefly introduce all the significant aspects of microbiology.

MICROBIOLOGY

Microbiology is the study of microorganisms, microscopic organisms that include in particular the bacteria, a large group of very small cells that have enormous basic and practical significance. Microbiology considers all aspects of microbial cells, including their structure, metabolism, diversity, genetics and evolution, ecology and roles in infectious diseases. Microbiology is composed of several subdisciplines, each of which is focused on part of the broader science. Scientists who study microorganisms, called microbiologists, typically specialise in one or more of these areas.

The science of microbiology developed later than other biological sciences, primarily because needed tools, such as the microscope, had to be developed to convincingly prove that microorganisms (also called microbes) exist. Following early microscopic discoveries, methods for the culture and identification of microorganisms were developed, and from these, our understanding of the enormous beneficial and detrimental effects of microbes began to unfold.

Classification and Basic Characteristics of Microorganisms

Microorganisms encompass an enormous diversity of microscopic life forms, each with distinct characteristics. On the basis of their genotypic (genetic) and phenotypic (observed) properties, all organisms are classified into one of three domains – the Bacteria, Archaea or Eukarya – and numerous examples of microorganisms are found in all three. The comparison of ribosomal ribonucleic acid (rRNA) gene sequences has been especially important in determining the evolutionary, or phylogenetic, relationships of organisms.

Although very different on a phylogenetic level, the Bacteria and Archaea (traditionally called prokaryotes) are structurally similar in that cells of these groups do not typically contain membrane-bound organelles and, therefore, show a lesser degree of cellular compartmentalisation than organisms belonging to the Eukarya (the eukaryotes). The presence of membrane-bound organelles, including a defined nucleus, is the hallmark of eukaryotic cells, and microbial Eukarya include fungi, protists and certain helminths, especially the parasitic worms.

The Bacteria and Archaea

The Bacteria and Archaea are vast groups of microorganisms consisting of potentially hundreds of thousands to millions of species, most of which remain uncharacterised. These microbes are ubiquitous, inhabiting and subsist in nearly every imaginable environment on Earth. Various species thrive on or within every plant and animal, within and underneath massive glaciers, in hypersaline waters of the Great Salt Lake and the Dead Sea, and even in boiling hot springs and deep sea volcanic (hydrothermal) vents. Many microorganisms, called extremophiles, are able to thrive in environmental conditions that humans would consider punishingly harsh, and numerous members of the Archaea, in particular, excel in this lifestyle.

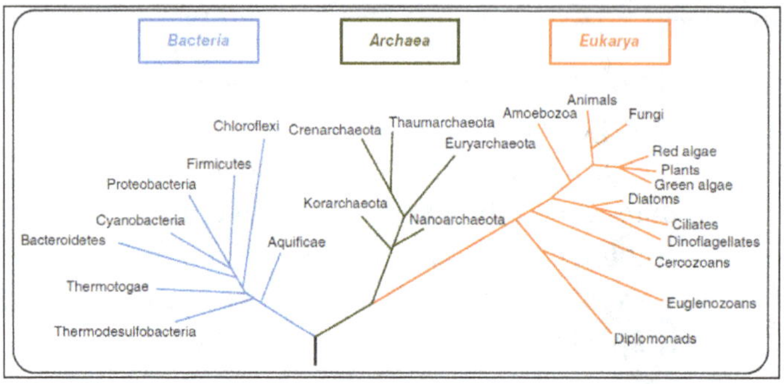

Universal phylogenetic tree showing relationships between major lineages of the three domains of life (Bacteria, Archaea and Eukarya). Tree topology and branch lengths were determined by comparative small subunit (SSU) rRNA gene sequence analysis.

Bacteria and Archaea are classified based on (1) phylogenetic distinctions, (2) structural and morphological characteristics, such as cell shape, size and arrangement and (3) biochemical and physiological traits, such as growth factor requirements, range of carbon and energy sources and end products of metabolism. For example, the Cyanobacteria are a large phylum of physiologically and phylogenetically related Bacteria that carry out oxygenic (O_2-producing) photosynthesis using chlorophyll-based pigments.

The classification of Bacteria and Archaea into various taxonomic groups is rapidly undergoing revision and modification as novel species continue to be discovered and as insights from the field of genomics provide more accurate information of phylogenetic relationships. Molecular tools have become essential for classifying Bacteria and Archaea that resist laboratory culture and thus can only be identified by analysing sequences of DNA or RNA isolated from natural samples.

Cells of Bacteria and Archaea exist in three major morphological forms: Spherical (coccus), rod-shaped (bacillus) and spiral-shaped (spirillum). Less common cell morphologies also exist, such as tightly coiled, appendaged and filamentous. All of these cell types are generally very small; most rods are 0.5–1 μm wide and 1–4 μm long, and a typical coccus has a diameter of 0.6–1 μm (about one-tenth the diameter of a human red blood cell). Many of these cells are motile by means of one or more rotating appendages called flagella. And unlike animal cells and protozoa, most species of Bacteria and Archaea have a cell wall, a strong layer of material located outside the cell membrane that confers structural integrity and shape to the cell.

Most species of Bacteria and Archaea divide by binary fission, a process in which one cell divides into two identical daughter cells following replication of the parental cell's DNA. Bacterial and archaeal genomes are usually organised into a single, circular chromosome, and many species also have small, extrachromosomal, circular pieces of DNA called plasmids, which often carry genes for antibiotic resistance, toxin production or specialised metabolisms. The chromosome is densely coiled in the cytoplasm to create a nucleoid, which is analogous to the nucleus of a eukaryotic cell but differs in that it is not bound by a membrane.

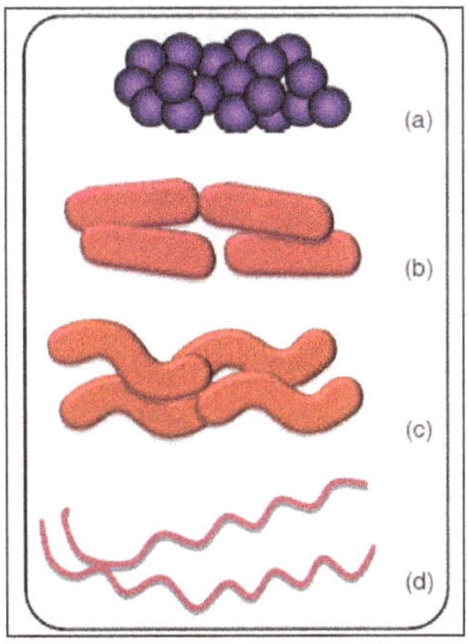

Major morphological forms of bacterial cells: (a) coccus (plural, cocci); (b) rod (bacillus; plural, bacilli); (c) spirillum (plural, spirilla) and (d) spirochete.

Microorganisms of the Domain Eukarya

Like Bacteria and Archaea, eukaryotic microorganisms are remarkably diverse, consisting of hundreds of thousands of species of fungi, protozoa and algae, as well as hundreds of species of parasitic worms. Fungi are a major component of soil ecosystems. Similar to plants, cells of fungi have rigid cell walls and are nonmotile, but unlike plants, they lack chlorophyll and are nonphotosynthetic. Instead, fungi subsist by degrading dead plant and animal matter, and therefore, along with bacteria, play a key role in the decomposition and recycling of nutrients.

Fungi exist in two basic forms: moulds, which consist of filaments called hyphae that can form into masses known as mycelia, and yeasts, which are unicellular and typically oval-shaped. Some fungi are dimorphic in that they can assume either morphology depending on environmental conditions. Fungi are capable of sexual reproduction or asexual reproduction, both of which may result in the production of spores that can germinate to form new hyphae. In addition, yeasts often reproduce asexually by budding, a process in which a new daughter cell develops on the surface of a parent cell before eventually breaking away.

Protozoa are unicellular, mostly nonphotosynthetic protists that lack cell walls. Some protozoa are large enough to be seen with the unaided eye, although most are microscopic. Many protozoa are capable of reproducing sexually or asexually. Depending on the species, asexual reproduction may occur through any of several mechanisms, including budding, spore formation or mitotic fission. A common mechanism of sexual reproduction in protozoa is conjugation, in which two cells join, exchange genetic material and produce progeny by budding or fission.

Protozoa are diverse in their habitats and distribution. Most protozoa obtain energy by breaking down ingested foods via aerobic respiration, but some species are anaerobes that lack mitochondria (the energy-generating organelle of eukaryotic cells) and instead obtain their energy through fermentation. Fermentative protozoa inhabit anoxic environments, such as the digestive tract of certain animals, where they often establish symbiotic relationships that may be either beneficial or harmful to their host.

Some protozoa cause devastating human diseases, such as malaria (caused by species of Plasmodium), while others (e.g. Paramecium) are innocuous members of the biosphere, where they exist as important components of the food chain.

Algae are plant-like protists that are distinguished from fungi and most protozoa by their ability to perform photosynthesis using chlorophyll pigments, and they comprise much of the basis of the food chain in marine and freshwater environments. Algae exhibit a variety of morphological forms, including unicellular, filamentous, colonial and large multicellular aggregates called kelps or 'seaweeds' that can attain lengths of up to 50 m. Some algae have become increasingly important sources of food or food additives for humans. For example, the red alga Porphyra, known as nori, is popular in sushi preparation, and other red algae are the source of agar, a polysaccharide used as a solidifying agent to make Petri plate culture media as or as a thickener for a variety of foods. Most algae reproduce asexually, whereas others form spores or reproduce by fragmentation of cells from larger aggregates. Some algae reproduce sexually by forming diploid zygotes from haploid gametes.

The helminths are a group of multicellular animals that includes roundworms and flatworms. Even though some of these organisms are visible to the naked eye, they are important topics of study formicrobiologists because (1) they often have microscopic larval forms as part of their life cycles and (2) many species of helminths are parasitic and cause important infectious diseases. Helminth life cycles can be complex and often require multiple hosts for the different stages. In addition to humans, these hosts may include other mammals, insects (e.g. flies and mosquitoes), various fish species and certain aquatic invertebrates, such as squid, snails and crustaceans.

A helminth infection usually begins with either an insect bite or an accidental ingestion of worm eggs or larvae. However, a few worm species, such as those that cause hookworm and schistosomiasis, are capable of burrowing directly through the skin.

The most important means of preventing parasitic worm infections include thoroughly cooking foods, drinking only purified or boiled water and employing repellents or physical barriers to prevent insect bites. Effective treatment of established infections is often possible using antihelminthic drugs.

Viruses are acellular microbes that require living host cells to multiply; thus, they are obligate intracellular parasites. Structurally, viruses are quite simple, often consisting of only DNA or RNA (the viral genome) surrounded by a simple protein coat having either a helical or icosahedral morphology. Viruses that infect bacteria, called bacteriophages, often have a complex morphology in that they exhibit a combination of these two forms. Most viruses are too small to be seen with even the best light microscopes, and because of their tiny size and dependence on host cells, their genomes are typically quite small, in some cases consisting of only two genes.

Viruses replicate within an infected cell by commandeering the host's enzymatic machinery, which may include use of the host cell's nucleic acid polymerases (enzymes that make DNA or RNA) and/or ribosomes. Following viral replication, progeny viruses are released, either by lysis of the host cell or by budding from the host cell's membrane. Therefore, in many cases, viral infections lead to death of the host cell, either abruptly or eventually. Viruses cause many serious diseases, including acquired immunodeficiency syndrome (AIDS), influenza, measles, poliomyelitis, rabies and haemorrhagic fevers, such as Ebola.

With a broad overview of microbial diversity in place, we can now explore the crucial role that microorganisms play in the environment around us.

Microbial Ecology

Microbial ecology is focused on how microbial communities interact with each other and their environments. A microbial community is an assembly of one or more populations of cells, each population composed of a single kind or species of microbe. An ecosystem is a dynamic complex of living organisms and their abiotic surroundings, all of which interact as a functional unit.

In terrestrial and aquatic ecosystems, microorganisms interact with each other and with the plants and animals in the ecosystem. Microbes play essential roles in these ecosystems by cycling inorganic nutrients and both producing and consuming organic matter. Associations of specific microbes with specific plants or animals are also quite common, and many of these associations are essential for the health and well-being of the plant or animal.

Soil, Water and Higher Organisms as Homes for Microbial Communities

Soil is the loose outer material that comprises much of Earth's surface and forms over long periods of time from a combination of biological and chemical processes. Soils often contain large numbers of microorganisms, and depending on the amount of organic matter present and the soil pH, salinity, degree of aeration and other abiotic factors, soil microbial communities can be relatively simple or highly complex. In addition to soil, however, microbial communities also exist deep underground in the Earth's subsurface, fed by nutrients transported by groundwater.

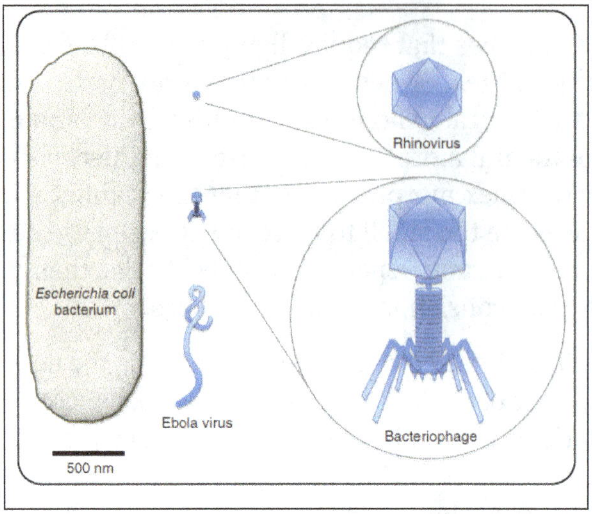

Major morphological forms of viruses. Most viruses are considerably smaller than cells (a representation of an Escherichia coli cell is shown for size comparison). Viruses of eukaryotes typically have either icosahedral (e.g. rhinoviruses) or helical (e.g. ebola virus) symmetry, whereas viruses that infect bacteria (bacteriophages) often exhibit a complex 'head-and-tail' combination of these forms.

In soils, microbial numbers are typically greatest in and around plant roots, a zone called the rhizosphere. It is here that organic matter excreted from the roots and from dead plant material greatly stimulates the activities of microbial communities. Temporal changes in the abundance and composition of soil microbial communities occur from variations in moisture, organic matter inputs and temperature. In contrast to surface soil, subsurface microbial communities are less dynamic due to more predictable conditions. Various Bacteria, Archaea, microbial eukaryotes and viruses inhabit soils and the deep subsurface, and many important nutrient cycling reactions occur there, including major transformations of the elements C, N and S, key constituents of living organisms.

Aerobic bacteria and fungi in soils consume oxygen in their respiratory activities. Many bacteria are anaerobes, carrying out various types of fermentation or anaerobic respiration (a form of respiration in which an oxidant other than O_2 is used). However, in the final analysis, aerobic and anaerobic respiration and fermentation all accomplish the same thing; these metabolisms oxidise organic carbon, returning it to CO_2. Many anoxic (O_2-free) zones exist in soils and thus aerobes and anaerobes coexist there.

Aquatic microbial ecosystems include both freshwaters and ocean waters. These two environments differ in many ways, including salinity, average temperature, depth and nutrient levels. Photosynthetic microbial communities play important roles in aquatic ecosystems. The CO_2-fixing (autotrophic) activities of these organisms provide not only the organic carbon needed for their own metabolism and growth but also the organic matter and oxygen needed by heterotrophic microbes present in the aquatic microbial community. Photosynthetic microbes are indeed the base of the aquatic food chain and thus are critical components of aquatic ecosystems.

The sediments of fresh and marine waters are hotbeds of distinctly different anaerobic metabolisms. In freshwater sediments, the bulk of organic carbon is eventually degraded to CH_4 (methane). Methane is formed by methanogenic Archaea that reduce CO_2 to CH_4 using hydrogen (H_2) as reductant in the anaerobic respiration called methanogenesis. In contrast to freshwater sediments, marine sediments contain large amounts of salts, including sulphate (SO_4^{2-}). Sulphate respiration, whereby SO_4^{2-} is reduced to H_2S, is the dominant form of anaerobic respiration in marine sediments because it is more energetically favourable than methanogenesis (the low levels of SO_4^{2-} in most freshwater sediments limits sulphate reduction and favours methanogenesis).

Many microorganisms form relationships with other organisms, which can include other microbes, plants or animals. Such associations are called symbioses (literally, 'living together'). Plants interact with microorganisms through their roots and leaf surfaces, but in some cases, the association becomes highly specific and intimate, including actual growth within the plant tissue.

Such is the case with the legume–root nodule symbiosis, an association in which tumour-like nodules form on the roots of leguminous plants (plants that bear their seeds in pods), such as soybeans and peanuts. The root nodules provide a habitat for bacteria that fix atmospheric nitrogen into ammonia (N_2 + 6H→ $2NH_3$). The plant then uses the ammonia as a source of nitrogen to make proteins and nucleic acids, allowing it to thrive in nitrogen-poor soils.

Animal–microbe symbioses are also quite common. The human large intestine, for example, contains enormous numbers of bacterial cells that form a huge microbial community, the human gut microbiome, the composition of which varies from person to person and is influenced by diet, health and other factors. As more is learned about the microbiomes of different animals, it has become clear that any perturbations in this species-specific microbial community can affect an animal's physical condition and susceptibility to disease. In some cases, the microbiome is essential for the very nutrition of the animal. For example, in the rumen (forestomach) of animals such as cows and sheep, bacteria and protozoa digest cellulose and ferment the released glucose to fatty acids, which are taken up by the animal. This allows the ruminant to subsist on a diet of plant matter, which is primarily cellulose.

Microbial Nutrient Cycling

The key nutrients for life are cycled by both microorganisms and by plants and animals, but for any given nutrient, it is microbial activities that dominate. The major nutrient cycles include those of carbon, nitrogen and sulphur. Carbon is cycled primarily through CO_2 (carbon dioxide) and the large pool of organic compounds present in living organisms. CO_2 is reduced to organic compounds by plants but also by many different microorganisms, both photosynthetic and chemolithotrophic. Organic matter, either excreted by living organisms or released from dead organisms, is eventually oxidized to CO_2 or is converted to methane (CH_4) by methanogenic archaea and later oxidised to CO_2 by methane-consuming bacteria.

Nitrogen compounds can be either oxidised or reduced by microbes depending on the compound. Ammonia (NH_3) is oxidized to nitrate (NO_3^-) by a group of chemolithotrophs called the nitrifying bacteria. Nitrate is then reduced to atmospheric nitrogen (N_2) by nitrate-respiring anaerobic bacteria (denitrifiers) or to NH_3 by ammonifying bacteria. The major remaining link in the nitrogen cycle is the reduction of N_2 to NH_3 by the nitrogen-fixing bacteria. This process is an important means of enriching soils in usable nitrogen and is the key to the soybean–root nodule symbiosis.

Sulphur is cycled primarily between SO_4^{2-} and sulphide (H_2S) through the activities of sulphate-reducing (sulphate-respiring) bacteria and chemolithotrophic sulphide-oxidising bacteria. Elemental sulphur (S_0) is often an intermediate product in these metabolisms and can be either oxidised to sulphate or reduced to sulphide. Sulphur-oxidising bacteria are primarily aerobes and participate in the sulphur cycle by consuming sulphide – a toxic substance for higher organisms – and generating sulphate, a key plant nutrient. By contrast, sulphate-reducing bacteria are anaerobes and oxidise organic matter to CO_2 in anoxic environments, generating sulphide. Some photosynthetic bacteria can also oxidise sulphide to support autotrophic growth ($CO_2 + H_2S \rightarrow \text{Organicmatter} + SO_4^{2-}$). Other key elements, such as P and Fe, are also cycled in nature, and like the C, N and S cycles, the cycling of these nutrients is driven primarily by the activities of microorganisms.

Medical Microbiology

While the vast majority of microorganisms provide beneficial services to humans, a few species (<1%), called pathogens, are potentially harmful. When pathogens successfully invade, multiply and cause damage to the host, disease ensues. Pathogens can damage the host in two ways: toxicity and invasiveness. Several pathogens produce toxins that can harm or even be lethal to humans. Some toxins, called exotoxins, are proteins secreted by the pathogen once established in the host. By contrast, endotoxins make up part of the outer layer (cell wall) of the cell itself. Bacterial exotoxins are among the most potent toxins known.

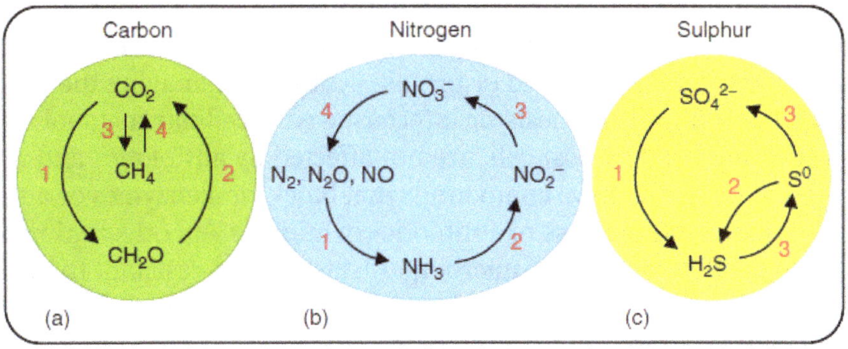

Key elemental cycles driven predominantly by the activity of microorganisms. (a) Carbon is cycled between inorganic (CO_2) and organic (CH_2O) forms by the actions of autotrophs (1) and heterotrophs (2). In addition,methane (CH_4) is produced (3) or consumed (4) by methanogens or methanotrophs, respectively. (b) Nitrogen compounds are cycled by nitrogen-fixing (1), nitrosifying (2), nitrifying (3) and denitrifying (4) bacteria. (c) In the sulphur cycle, sulphide is produced by sulphate- and sulphur-reducing bacteria (1 and 2, respectively). Sulphur chemolithotrophic bacteria oxidise sulphide to sulphate through an elemental sulphur intermediate.

For example, just 1 µg (one-millionth of a gram) of botulinum exotoxin can kill an adult human. Invasive pathogens harm the host by directly damaging tissues or depriving them of nutrients. Bacterial invasion begins with attachment to host cells, which is facilitated by cellular appendages, such as fimbriae or pili. Tissue destruction occurs through the secretion of enzymes, such as leukocidins, which destroy white blood cells, and collagenases, which erode collagen, a fibrous protein found in various connective tissues. Some bacteria produce dense polymer coatings outside their cell walls called capsules that protect the invading pathogen from the host's immune response. Hosts respond to microbial invasion and the presence of toxins by producing protective antibodies that bind to and neutralise the microbe or toxin and by amassing white blood cells to attack invading pathogens by either consuming them or producing toxic agents to kill them.

Human hosts acquire infectious diseases by different routes, including the respiratory tract, the oral cavity and digestive system and the skin and genitourinary system. In some cases, pathogens are transmitted person-to-person, as in sexually transmitted infections (e.g. gonorrhoea and syphilis) or respiratory diseases (e.g. tuberculosis and influenza). In other instances, pathogens are transmitted indirectly through inanimate objects, such as clothing or towels, or through vectors, animals that carry pathogens between susceptible hosts. Lyme disease and malaria are examples of diseases that are transmitted indirectly to humans by arthropod vectors, the deer tick and Anopheles mosquito, respectively. After a pathogen is transmitted to a susceptible host, an incubation period occurs in which the pathogen multiplies and becomes established in the host before signs and symptoms of disease appear.

Antibiotics are antimicrobial chemicals produced by various microorganisms. Antibiotics are often administered to help fight infectious microorganisms and kill or inhibit pathogens by disrupting nucleic acid or protein synthesis, damaging the plasma membrane, preventing cell wall synthesis or interfering with cellular metabolism. Viruses, which replicate only inside a host cell, are not affected by antibiotics and must therefore be treated by other chemical compounds that block viral enzymes or alter viral nucleic acids. The indiscriminate use of antibiotics in humans and other animals in recent years has led to the development of microorganisms that are resistant to many of these drugs. These antibiotic-resistant microbes challenge conventional chemotherapy and are of major concern to medical professionals.

An alternative strategy to combating microbial infections is to employ vaccines to prevent pathogens from becoming established in the body in the first place. Vaccines induce an immune response in the host, and this triggers the production of protective antibodies that provide immunity against specific infectious agents and other foreign antigens. Vaccines have been particularly helpful in preventing common childhood diseases, including diphtheria, whooping cough (pertussis), poliomyelitis, measles and mumps.

Applications in Microbiology

Commercial Products from Microorganisms

Microorganisms can be harnessed to make many valuable products, and industrial microbiology and biotechnology are the subdisciplines of microbiology focused on these tasks. Industrial microbiology uses microorganisms to synthesise products in large amounts. This is done by taking microbes that naturally produce some substance of relatively low value – for example, an antibiotic or alcohol – and selecting for 'overproducing strains' that can be grown on a huge scale; the resulting product may be made by tons or thousands of litres. Biotechnology, by contrast, employs genetically engineered microbes to synthesise small amounts of very high-value products that the microbes are otherwise unable to make, such as a human protein. Some major products of industrial microbiology and biotechnology are listed in table.

Table: Selected products of industrial microbiology and biotechnology.

Product	Example
Industrial microbiology	
Antibiotics	Penicillin, tetracycline
Enzymes	Laundry proteases and lipases, glucose isomerase
Food additives	Vitamins, amino acids, citric acid
Alcohol/chemicals	Ethanol in alcoholic beverages and gasohol, butanol, biodiesel and steroids

Biotechnology	
Human hormones	Insulin, somatotropin
Blood proteins	Tissue plasminogen activator
Immune modulators	Interferons, tumour necrosis factor
Therapeutic enzymes	DNase

Food and Mining Microbiology

Food production and mining are obviously quite different activities but both owe their success to the microbial world. Certain microbes are used in the preparation of common food products while others are used in the mining industry to extract valuable minerals from crude ores.

Yeast is the key catalyst in the production of baked goods and alcoholic beverages. The fermentative metabolism of yeasts (glucose → 2 ethanol + 2 CO_2) generates key products for the baker (CO_2 to raise the dough) and the brewer (ethyl alcohol). Many cheeses owe their characteristic flavours to the activities of fermentative microbes. For example, the key components in Swiss cheese – propionic acid and CO_2 – are responsible for the distinctive nutty taste and characteristic holes (eyes), respectively, of this popular cheese. Other major food products of microbial origin include fermented milks (e.g. buttermilk and yogurt), meats (e.g. salami and summer sausage) and vegetables (e.g. sauerkraut and pickles).

The mining of copper, gold and a few other metals relies on the leaching properties of bacteria. These valuable metals are typically present in low amounts in complex mineral ores that contain large amounts of iron. The metabolic activities of various acid-tolerant, iron-oxidising bacteria release these metals from the ores and generate acidic solutions (the leachate) from which the now solubilised metals can be concentrated and harvested by chemical treatment. Copper, widely used in electrical equipment, piping and the brewing industry, is the most commonly leached metal. The strategic element uranium is also mined by microbial leaching processes.

Environmental Applications of Microorganisms

Microorganisms play critically important roles in the environment. We have already considered the key roles that microbes play in nature's major nutrient cycles. But in addition to these more or less continuous activities, microorganisms have been exploited for purifying wastewaters and for cleaning pollution in the environment, a process called bioremediation.

Sewage and other wastewaters must be treated before they can be released into natural waterways. This is because without treatment, the massive influx of organic matter and mineral nutrients would trigger extensive microbial growth and O_2 consumption, causing die-offs of plants and animals and diminishing the aesthetic and recreational value of the water. To deal with the high nutrient load of wastewaters, elaborate treatment facilities are used to stimulate the activities of complex microbial communities to remove as much organic carbon and other polluting nutrients (such as nitrates and phosphates) from the wastewaters as possible. Following treatment, the water can then be safely released into rivers or other bodies of water.

To produce potable drinking water, additional treatment is necessary to remove as many potentially pathogenic microorganisms and remaining toxic substances as possible. Drinking water production includes the coagulation and filtration of already high-quality surface or subsurface waters followed by disinfection with chlorine and transport of the water through water mains to the consumer. The entire process of drinking water production must be carefully performed and monitored to prevent breakdowns that can lead to incidents of serious waterborne illness, such as cholera or typhoid fever.

When pollution of the environment occurs, either from natural events or from the activities of humans, microorganisms can be harnessed to clean up the mess. Microbial bioremediation is typically the most cost-effective method of removing environmental pollutants and, in many cases; it is the only practical way to accomplish the job. Bioremediation is grounded in the astounding diversity of metabolic reactions capable in the microbial world.

Thus, if some pollutant, such as crude oil, is spilled in the environment, oil-consuming microbes applied to the spill site can clean up themess by oxidising hydrocarbons in the oil to CO_2.

In a similar manner, microbes that can degrade pesticides, such as insecticides and herbicides, are beneficial in keeping these poisonous substances from accumulating in the environment and damaging plants and animals that were not the original targets of these agents. Although not every substance that humans have created is biodegradable (e.g. teflon is not), the vast majority of pollutants are, and it is through the activities of microorganisms that these undesirable substances are converted into compounds that can enter the natural nutrient cycles.

Humans owe a considerable debt to the microbial world for keeping planet Earth habitable and healthy. If cyanobacteria had never become established on Earth, then the oxygen we breathe and depend on would never have been produced. And if it were not for microbes today, the everyday activities of humans would eventually damage the environment beyond its capacity to sustain human life. The microbial world is clearly the foundation of the biosphere, and thus the science of microbiology, which attempts to understand this unusual world.

MICROBIOME

A microbiome is the community of micro-organisms living together in a particular habitat. Humans, animals and plants have their own unique microbiomes, but so do soils, oceans and even buildings.

There has been an explosion in our understanding of the human microbiome (the genome of all our microbes) in the recent years. Advances in genome sequencing technologies and metagenomic analysis (genetic study of genomes taken directly from environmental samples) have enabled scientists to study these microbes and their function and to research microbiome–host interactions both in health and disease. The human microbiome has an estimated 100 trillion microbes, the bulk of which live in our gut.

The Human Microbiome: An Organ in its Own Right

The human microbiome is composed of communities of bacteria (and viruses and fungi) that have a greater complexity than the human genome itself. Large-scale metagenomic projects (community and environmental genomics), such as the European Metagenomics of the Human Intestinal Tract and the Human Microbiome Project, have reported 3.3 million unique protein-encoding genes as compared with the entire human genome, which has around 23 000 genes. These studies have described the beneficial functions of the normal gut microbiota on health down to the genetic level. The human microbiome has extensive functions such as development of immunity, defence against pathogens, host nutrition including production of short-chain fatty acids important in host energy metabolism, synthesis of vitamins and fat storage as well as an influence on human behaviour, making it an essential organ of the body without which we would not function correctly.

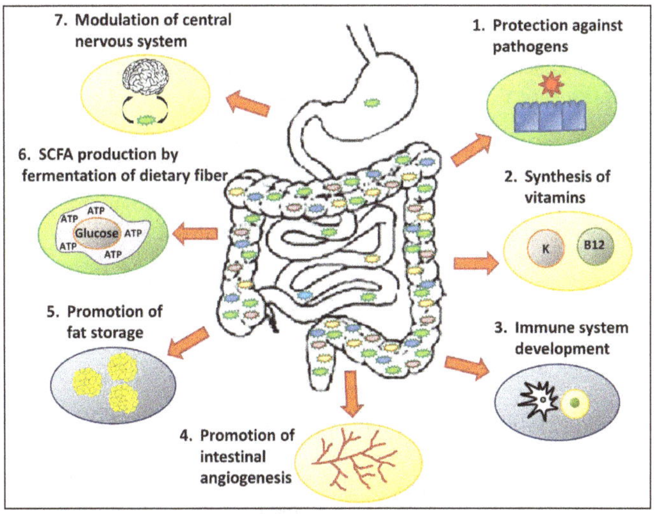

The human microbiome plays an important role in control of vital homeostatic mechanisms in the body. These include enhanced metabolism, resistance to infection and inflammation, prevention against autoimmunity as well as an effect on the gut–brain axis. SCFA, short-chain fatty acid.

Microbiome Composition

Unlike the host genome, which is relatively constant, the microbiome is dynamic and changes with early development, environmental factors such as diet and use of antibiotics and especially in response to disease. The most dramatic changes in composition occur in infancy and early childhood. The intestinal microbiome of an infant is affected by gestational age (full term or premature), mode of delivery (vaginal birth or caesarean section), type of feed (breast milk or formula feeds), maternal nutritional status (overweight or undernourished) and use of antibiotics. The complexity and plasticity of the infant microbiota during this early life development is believed to be important in maintaining homeostasis with the host's immune system.

A healthy human gut can house at least 1000 different species of bacteria, comprising of two major phyla, namely Bacteroidetes and Firmicutes. Elucidating the microbiome composition in healthy individuals is important to understanding what consequences changes in microbiome composition may have to human health and disease susceptibilities.

Other microbes also live in the gut including viruses, fungi and archaea. However, research on them has been small compared with that on bacteria.

The Human Microbiome and Disease

The relationship between changes in microbiome composition and disease pathogenesis is uncertain. The challenge is to identify whether microbial imbalance is related to disease, sometimes termed 'dysbiosis', and to be able to distinguish between cause and effect.

Table: Alterations in gut microbiota exert an influence on organ systems in the body leading to disease.

Organ	Examples of diseases linked with altered microbiota	Microbiota-mediated changes	References
Brain	Autism spectrum disorder	Abundance of bacterial toxins Disrupted fermentation	9 , 25
Lung	Asthma Cystic fibrosis	Reduced immunological tolerance Altered gene expression	11 , 12 , 21 , 23 ,24
Heart	Cardiovascular disease	Production of proinflammatory metabolites	13
Pancreas	Type 1 and type 2 diabetes	Reduced insulin sensitivity	8
Liver	Non-alcoholic fatty liver disease	Altered bile acid metabolism	10
Adipose tissue	Metabolic syndrome Obesity	Reduced intestinal gluconeogenesis Insulin resistance	8 , 14 , 15
Gastro-intestinal tract	Inflammatory bowel syndrome Irritable bowel syndrome Gut infections	Dysregulated immune response Altered mucosal barrier	17–20 , 22
Skin	Acne Eczema Allergic diseases	Increased pathogenic strains Dysregulated immune response	16

Inflammatory bowel disease (IBD)—This is one of the most extensively studied human conditions associated with the gut microbiota. The composition of the gut microbiota differs between healthy individuals and patients with IBD both in terms of species richness (ie, numbers of bacterial species) and species abundances (ie, number of individuals per species). As bacteria are identified by sequencing, rather than by functional characteristics in the culture laboratory, the individual bacterial species or genus (depending on the classification of sequence data) are commonly referred to as operational taxonomic units in microbiota research.

Studies have reported patients with IBD to have decreased bacterial diversity, and reduced abundances of Firmicutes and Bacteroidetes and an increased abundance of Proteobacteria compared with healthy individuals. Latest research suggests that IBD pathogenesis is due to the interaction of environmental factors (eg, smoking, diet and stress) and the host's genetic susceptibility, which is influenced by commensal microbiota, which activates either pathogenic or protective immune responses. Evidence from mouse models provides further support for the role of gut microbiota in pathogenesis of IBD.

Necrotising enterocolitis (NEC): The pathogenesis of NEC is multifactorial although the gut microbiota is thought to play a crucial role. Studies in both humans and animal models have described changes in the gut microbiota including a reduction in bacterial diversity and increased abundances of Proteobacteria in preterm infants with NEC compared with healthy preterm infants. However, the results have been inconsistent across studies, and to date, no single causative set of microorganisms has been identified.

Atopic diseases: Conditions such as eczema, asthma and food allergies are increasing in incidence. This is often linked to the hygiene hypothesis. It is thought that the lack of early-life exposure to microbial antigens in hygienic developed countries alters the microbiota composition of the infant gut, which disrupts immune development causing allergic disease. For example, species like Bacteroides fragilis reportedly induces immunological tolerance through immune receptor signalling pathways. Also the infant gut microbiota is affected by environmental factors including pets, residing in rural homes and siblings shown to have protective effects against asthma and allergies. The concept that altered microbiome composition influences childhood allergic disease susceptibility is further supported by data from epidemiological studies that report higher prevalence of atopic diseases in infants delivered by caesarean sections, formula fed infants and those exposed to antibiotics.

Type 1 diabetes: The gut microbiota is involved in regulation of the metabolic–immune axis.8 Research studies speculating the specific causative microbial composition and function have not been consistent. However, Bifidobacteria is believed to be protective, while Proteobacteria is a reported risk factor. Similarly, changes in gut microbiota caused by lifestyle (eg, mode of delivery and diet) are known risk factors for the development of type 1 diabetes, like birth by caesarean section reportedly increases the risk of developing type 1 diabetes by 20%.7

Autistic spectrum disorder (ASD): The gut microbiota can influence human behaviour by modulating the gut–brain axis via endocrine (cortisol), immune (cytokines) and neural (vagus and enteric nervous system) signalling pathways. Several studies have described an altered gut microbiota in children with ASD compared with developmentally normal children. However, more work is needed to establish if this is important in the pathogenesis of the condition.

Future Potential of Microbiota Manipulation for the Treatment of Disease

Manipulation of the microbiota as a therapeutic tool is a rapidly advancing field in microbiome research. There is an abundance of data suggesting treatments capable of reversing dysbiosis are effective in managing certain human diseases. Targeted antibiotic use to eliminate select microbiota, probiotics and prebiotics to encourage the expansion of beneficial bacteria and faecal microbiota transplantation to restore bacterial communities are some of the approaches that are currently under investigation or in use. These studies will provide a greater understanding of the host–microbiome interactions that impact on disease.

As paediatricians, we need to keep abreast of the research as within the next decade, it is highly likely that we may be using such treatments in our practice.

TYPES OF MICROBIAL MUTANTS

Auxotrophic Mutants

Auxotrophic mutants are those that are nutritionally deficient, i.e., they are unable to synthesize essential metabolites or growth factors (amino acid, purine, pyrimidine, vitamins, etc.) that are synthesized de novo by the parental or wild-type strains (prototroph) of the species.

An auxotrophic mutant will not grow when the essential metabolite (growth factor) is absent in culture medium; it grows and reproduces only when the required metabolite is supplied in the medium.

Resistant Mutants

Resistant mutants exhibit an increased tolerance to inhibitory agents, particularly antibiotics and phages. Microbes may develop such mutants spontaneously through a range of mechanisms, e.g., acquisition of enzymes able to metabolize antibiotics, and loss of cell surface components which act as phage-receptors.

Cryptic Mutants

Cryptic mutants are those that have lost a specific function but retain the intracellular activities to catalyze the reactions of the function, for example, loss of a permease.

Conditionally Expressed Mutants

These mutants remain as wild-type phenotype under one set of conditions but may express as mutant phenotype when these conditions alter.

Antigenic Mutants

Antigenic mutants show a change in the surface structure and composition of the microbial cell. Some antigenic mutants exhibit altered colonial shape due to effect on cell wall synthesis, and others lose the capacity to make specific cell components, e.g., capsular polysaccharides.

Metabolic Mutants

These mutants exhibit altered metabolic ability particularly the fermentation- ability of decreased or increased capacity to produce some end-product.

Regulatory Mutants

Regulatory mutants are those in which mutation affects either the regulatory region of the promoter gene or the activity of a regulatory protein. Mutants of Bacillus subtilis are grossly deficient in the enzymes needed to separate daughter cell after division and they grow at normal rates as very long chains or un-separated cells.

Temperature-sensitive Mutants

Temperature-sensitive mutants grow at one temperature but not at another. Most of them are heat-sensitive, some, however, are cold-sensitive. The temperature sensitivity usually results from increasing heat or cold liability of the mutant gene-product.

For example, a temperature- sensitive mutant of E. coli fails to provide stability to enzymes involved in lactose utilization and does not grow on lactose at that temperature, while it retains the ability to grow on lactose at low temperature.

ISOLATION

In microbiology, the term isolation refers to the separation of a strain from a natural, mixed population of living microbes, as present in the environment, for example in water or soil flora, or from living beings with skin flora, oral flora or gut flora, in order to identify the microbe(s) of interest. Historically, the laboratory techniques of isolation first developed in the field of bacteriology and parasitology (during the 19th century), before those in virology during the 20th century. Methods of microbial isolation have drastically changed over the past 50 years, from a labor perspective with increasing mechanization, and in regard to the technology involved, and hence speed and accuracy.

General Techniques

In order to isolate a microbe from a natural, mixed population of living microbes, as present in the environment, for example in water or soil flora, or from living beings with skin flora, oral flora or gut flora, one has to separate it from the mix.

Traditionally microbes have been cultured in order to identify the microbe(s) of interest based on its growth characteristics. Depending on the expected density and viability of microbes present in a liquid sample, physical methods to increase the gradient as for example serial dilution or centrifugation may be chosen. In order to isolate organisms in materials with high microbial content, such as sewage, soil or stool, serial dilutions will increase the chance of separating a mixture.

In a liquid medium with few or no expected organisms, from an area that is normally sterile (such as CSF, blood inside the circulatory system) centrifugation, decanting the supernatant and using only the sediment will increase the chance to grow and isolate bacteria or the usually cell-associated viruses.

If one expects or looks for a particularly fastidious organism, the microbiological culture and isolation techniques will have to be geared towards that microbe. For example, a bacterium that dies when exposed to air, can only be isolated if the sample is carried and processed under airless or anaerobic conditions. A bacterium that dies when exposed to room temperature (thermophilic) requires a pre-warmed transport container, and a microbe that dries and dies when carried on a cotton swab will need a viral transport medium before it can be cultured successfully.

Bacterial and Fungal Culture

Inoculation

Laboratory technicians inoculate the sample onto certain solid agar plates with the streak plate method or into liquid culture medium, depending what the objective of the isolation is:

- If one wants to isolate only a particular group of bacteria, such as Group A Streptococcus from a throat swab, one can use a selective medium that will suppress the growth of concomitant bacteria expected in the mix (by antibiotics present in the agar), so that only Streptococci are "selected", i.e. visibly stand out. To isolate fungi, Sabouraud agar can be used. Alternatively, lethal conditions for streptococci and gram negative bacteria like high salt concentrations in Mannitol salt agar favor survival of any staphylococci present in a sample of gut bacteria, and phenol red in the agar acts as a ph indicator showing if the bacteria are able to ferment mannitol by excreting acid into the medium. In other agar substances are added to exploit an organism's ability to produce a visible pigment (e.g. granada medium for Group B Streptococcus) which changes the bacterial colony's color, or to dissolve blood agar by hemolysis so that they can be more easily spotted. Some bacteria like Legionella species require particular nutrients or toxin binding as in charcoal to grow and therefore media such as Buffered charcoal yeast extract agar must be used.

- If one wants to isolate as many or all strains possible, different nutrient media as well as enriched media, such as blood agar and chocolate agar and anaerobic culture media such as thioglycolate broth need to be inoculated. To enumerate the growth, bacteria can be suspended in molten agar before it becomes solid, and then poured into petri dishes, the so-called 'pour plate method' which is used in environmental microbiology and food microbiology (e.g. dairy testing) to establish the so-called 'aerobic plate count'.

Incubation

After the sample is inoculated into or onto the choice media, they are incubated under the appropriate atmospheric settings, such as aerobic, anaerobic or microaerophilic conditions or with added carbon dioxide (5%), at different temperature settings, for example 37 °C in an incubator or in a refrigerator for cold enrichment, under appropriate light, for example strictly without light wrapped in paper or in a dark bottle for scotochromogen mycobacteria, and for different lengths of time, because different bacteria grow at a different speed, varying from hours (Escherichia coli) to weeks (e.g.mycobacteria).

At regular, serial intervals laboratory technicians and microbiologists inspect the media for signs of visible growth and record it. The inspection again has to occur under conditions favoring the isolate's survival, i.e. in an 'anaerobic chamber' for anaerobe bacteria for example, and under conditions that do not threaten the person looking at the plates from being infected by a particularly infectious microbe, i.e. under a biological safety cabinet for Yersinia pestis (plague) or Bacillus anthracis (anthrax) for example.

Identification

When bacteria have visibly grown, they are often still mixed. The identification of a microbe depends upon the isolation of an individual colony, as biochemical testing of a microbe to determine its different physiological features depends on a pure culture. To make a subculture, one again works in aseptic technique in microbiology, lifting a single colony off the agar surface with a loop and streaks the material into the 4 quadrants of an agar plate or all over if the colony was singular and did not look mixed.

Gram staining the raw sample before incubation or staining freshly grown colony material helps to determine if a colony consists of uniformly appearing bacteria or is mixed, and the color, and shape of bacteria allow a first classification based on morphology. In clinical microbiology numerous other staining techniques for particular organisms are used (acid fast bacterial stain for mycobacteria). Immunological staining techniques, such as direct immunofluorescence have been developed for medically important pathogens that are slow growing (Auramine-rhodamine stain for mycobacteria) or difficult to grow (such as Legionella pneumophila species) and where the test result would alter standard management and empirical therapy.

Biochemical testing of bacteria involves a set of agars in vials to separate motile from non-motile bacteria. In 1970 a miniaturized version was developed, called the analytical profile index.

Successful identification via e.g. genome sequencing and genomics depends on pure cultures.

Bacteria and Culture-independent

While the most rapid method to identify bacteria is by sequencing their 16S rRNA gene, which has been PCR-amplified beforehand, this does method not require isolation. Since most bacteria cannot be grown with conventional methods (particularly environmental or soil bacteria) metagenomics or metatranscriptomics are used, shotgun sequencing or PCR directed sequencing of the genome. Sequencing with mass spectrometry as in Matrix-assisted laser desorption/ionization (MALDI-TOF MS) is used in the analysis of clinical specimens to look for pathogens. Whole genome sequencing is an option for a singular organism that cannot be sufficiently characterized for identification. Small DNA microarrays can also be used for identification.

TECHNIQUES USED FOR THE ISOLATION OF MICROBIAL MUTANTS

Direct Observation Technique

In some cases, a colony growing on an agar plate can easily be seen to be different from the normal parental type (wild-type). For example, if the parental strain is pigmented, the observation of non-pigmented colonies may indicate the presence of mutants. Indicators can also be incorporated into the medium to detect microorganisms with and without particular metabolic capabilities.

For instance, pH indicators can be used into the medium to detect the production of acidic products. The indication of acid production by one microbial strain and not by other of the same microorganisms growing under identical conditions would show the presence of a mutant.

Enrichment Technique

Enrichment technique is employed especially in isolating mutants resistant against phages, antibiotics, or toxic chemicals. Phage-resistant mutants can be isolated simply by plating the mutagenized, phenotypically expressed microbial population on plates containing phage particles.

Cells expressing the parental wild-type phenotype are killed; only phage-resistant mutants develop into colonies. Such colonies are isolated. Similarly, mutants resistant to an antibiotic or a toxic chemical can be isolated by plating the microbial population with the antibiotic or the chemical.

Replica-plating Technique

Replica-plating technique is often used to isolate nutritional mutants (auxotrophs) as well as various other type of mutants, e.g., antibiotic resistant mutants.

For convenience, if one wants to isolate nutritional mutants employing replica-plating techniques, he is required to follow the following steps:

- Bacterial cultures are diluted, and the cells are spread on the surface of semi-solid nutrient agar medium in a Petri dish (called "master plate"). The medium in the master plate is a complete medium i.e., containing all the nutritional components required by the bacterial population. After a sufficient incubation period, each bacterium produces a visible colony on the surface of the agar in the master plate.
- A piece of sterile velvet cloth is stretched over a cylindrical block of wood or metal that is slightly smaller in diameter than the Petri dishes used in the process.
- The master plate is now inverted and gently pressed onto sterile velvet. Since the fibres of the velvet act as fine inoculating needle, some cells from each colony of the master plate stick to the velvet.
- Other Petri dish (called "replica plate") is taken containing a minimal medium i.e., a medium deficient with specific nutritional component.
- The replica-plate is now inverted and gently pressed onto the velvet thus stamping the bacterial cells onto the surface of its minimal medium. The replica plate is identically oriented at the application on the velvet with respect to mark placed on its rim so that the colonies that appear on the replica plate after incubation occupy positions congruent with those of their siblings on the master plate.
- After sufficient incubation, it is observed that a colony that develops on the complete medium of the master plate fails to develop on the minimal medium of the master plate that lacks a specific nutritional component. Such colony is marked on the master plate and is isolated; it represents mutant for that specific nutritional component not used in the minimal medium of the replica plate.

The replica-plating technique was developed by Joshua and Esther Lederberg in 1952 in order to provide direct evidence for the existence of pre-existing mutations originated spontaneously in microorganisms.

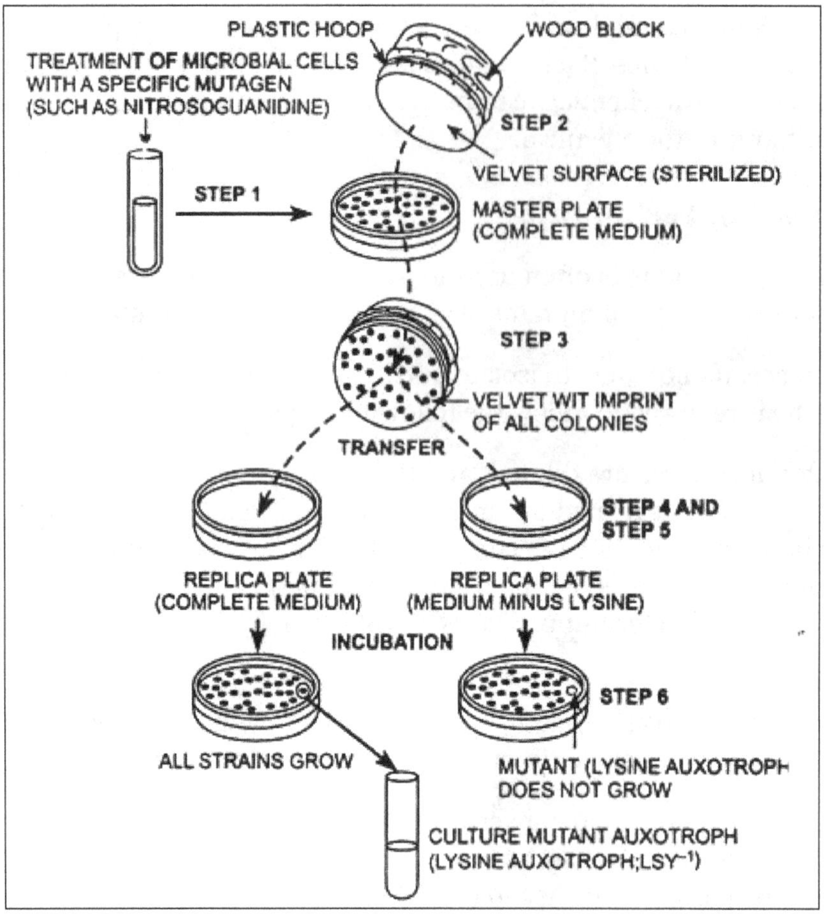

The Ames Test

This test was developed by Ames and coworkers and is based on histidine-requiring (his–) auxotrophic mutants of Salmonella typhimurium. Different his– mutants carry different types of mutations, i.e., transitions, transversions and frame- shifts.

In the Ames test, the frequency of reversion to his+ (prototrophy) is scored in the especially constructed his– mutants. This is done by placing a known number of mutant cells on medium lacking histidine and scoring the number of colonies formed. The frequency of cells forming colonies gives the frequency of reversion. The frequency of spontaneous reversion to his+ is quite rare, i.e., 10-8.

Ames test is routinely used to investigate the mutagenicity of various chemicals. Some of the chemicals may become mutagenic only when they are acted upon by liver enzymes.

For example, nitrates themselves are neither mutagenic nor carcinogenic. But in eukaryotic cells, nitrates are converted to introsamines, which are highly mutagenic and carcinogenic. In addition; some chemicals may be mutagenic only to replicating DNA.

The routine Ames test addresses to both these needs as follows:

1. The his-cells are plated onto a medium that contains traces of histidine, which is enough to allow a few cell divisions, but inadequate for visible colony formation.

2. The test chemical is incubated with rat liver extract containing the liver enzymes, i.e., the microsomal fraction. This allows modification of the chemical in the same way as it would be in the liver of animals.

The procedure of Ames test is as follows. The his– bacterial cells are incubated with the liver extract, and then plated onto a medium containing traces of histidine; this serves as the control plate.

The test plates contain the same medium but the his– cells are not treated with liver extract. The test chemical is treated with the rat liver extract and a filter paper disc is soaked in this solution. The filter paper disc is placed onto the medium of test plate.

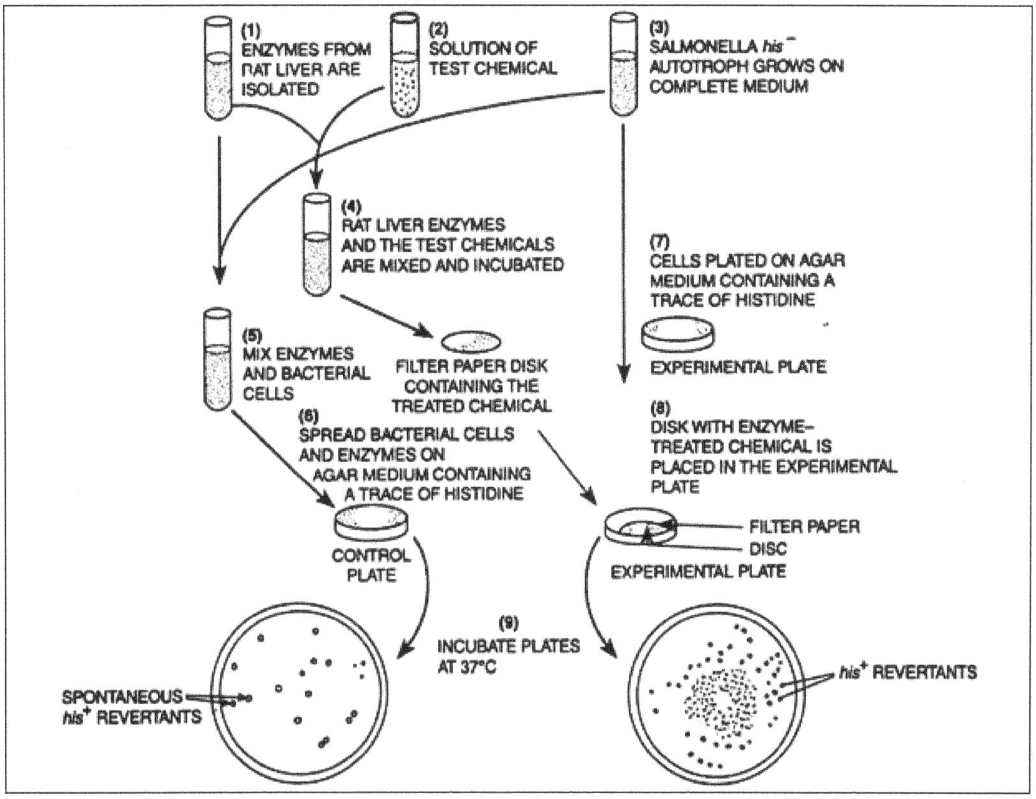

The chemical present in the filter paper acts on the his– cells growing in the test plate. The frequency of colonies formed in the control plate and the test plate are compared. An increase in the frequency in the case of test plate will indicate the test chemical to be mutagenic.

In order to increase the efficiency of the test, the his– strains used in the test are defective in DNA repair, and have increased permeability to chemicals. It has been observed that more than 90% of the chemicals that are mutagenic are also carcinogenic.

Subfields of Microbiology

CHAPTER 2

Microbiology is divided into a number of Subfields namely mycology, bacteriology, protozoology, virology, microbial genetics, cellular microbiology, food microbiology, industrial microbiology and soil microbiology. This chapter has been carefully written to provide an easy understanding of these different branches of microbiology.

Bacteria are the most abundant, versatile and oldest organisms on earth. They play vital roles in the world's ecology and in human industry and medicine. In this work we present an introduction to some of the basic principles of bacteriology. We describe the basic cell structure of bacteria and how it differs from the structure of eukaryotic cells. Basic concepts of bacterial growth in vitro and pure culture techniques are introduced. The various classes of antibiotics, their mechanisms of action and the nature of bacterial resistance are discussed. We describe the various mechanisms of gene transfer in bacteria including the manipulation of DNA in vitro.

Eubacteria

One current view of life on Earth classifies all cellular life forms into one of two groups, the eukaryotes and the prokaryotes. These terms are based upon the presence of a nuclear membrane, or "true nucleus," in eukaryotes and the absence of the same in prokaryotes. Most of the organisms that are most familiar to us (plants, animals and fungi) are eukaryotes. At the cellular level, eukaryotic organisms posses subcellular organelles such as mitochondria, golgi bodies an endoplasmic reticulum and in the case of green plants, chloroplasts. Eukaryotes have large (80S) ribosomes containing three ribosomal RNAs (rRNA) referred to as 28S, 18S and 5s rRNAs. Eukaryotes typically have sterol-containing membranes and diploid genomes.

The prokaryotes are divided into two kingdoms, the eubacteria and the archaea. The later, once referred to as the archaebacteria, are unicellular organism that morphologically resemble the true bacteria but phylogenetically are as closely related to eukaryotes as to eubacteria. The archaea were initially defined by their 16S ribosomal RNA sequences, but since have been found to have unique cellular lipids, cell envelopes and unique transcription and translation machinery. The cell walls of archaea lack peptidoglycan. The archaea include organisms that produce methane (methanogens), those that tolerate extremely high salt concentrations (halophiles) and those that grow under high temperature and low pH conditions (thermoacidophiles). The eubacteria are differentiated from the archaea.

Eubacterial cells are relatively small, ranging from 0.3 to 5 μm in diameter but typically about 1.0 μm in diameter. While they lack a nuclear membrane, their single circular DNA chromosome is within a specialized region of the cell called a nucleoid. They have smaller (70S) ribosomes containing 23S 16S and 5S rRNAs. Most eubacteria lack sterols in their cell membranes and produce cell walls containing peptidoglycan.

Importance of Bacteria

Bacteria are an extremely diverse group of organisms that play critical roles in global ecology and the cycling of essential elements like carbon, nitrogen, sulfur, phosphorous and oxygen. Bacteria can be found in everyplace on the planet that contains liquid water. Bacteria have been used for thousands of years to produce fermented products such as cheese, yogurt, sourdough bread, sauerkraut, poi, and many others. More recently, bacteria have been utilized for the industrial production of organic solvents, antibiotics and amino acids. With the advent of recombinant DNA techniques bacteria have been engineered to produce proteins for medicinal and industrial use. Naturally occurring bacteria colonize our skin, oral cavity and gastrointestinal tract often protecting us from pathogenic microbes. The study of bacteria has greatly increased our knowledge of how cells function and the understanding of the molecular basis of life.

Bacterial Morphology and the Cell Envelope

Shapes and Arrangements of Bacterial Cells

If one were given a culture of bacteria and asked to identify the organism the first test one would almost always perform would be a Gram stain. The Gram stain involves staining a thin layer of bacteria dried onto a microscope slide with two dyes, crystal violet and iodine. The stained area is then washed with a decolorizer containing acetone then stained again with safranin. If the stained organism has a cell wall containing a thick layer of peptidoglycan, the crystal violet stain will be retained and the cells will appear purple under the microscope; this would be considered a positive test and the organism would be classified as a Gram-positive organism. Alternatively, if the organism has a thin peptidoglycan layer, the crystal violet-iodine stain will be lost during the acetone wash step and the cells will be stained red with safranin. Thus Gram negative organism will appear red. The Gram stain permits discrimination of the two major types of bacteria, Gram negatives and Gram positives.

Microscopic examination allows further classification of an organism based upon cellular morphology and the arrangement of cells relative to each other. Although there are many variations, bacteria are usually classified as one of three basic shapes. A coccus, or cocci if plural, is a spherical shaped cell. A bacillus, or bacilli, is a rod shaped bacterium and a spirillum (spirilla) is helical shaped. The arrangements or groupings of cells include pairs, chains, clusters, tetrads, sarcina and palisades. The morphologies and groupings of some well-known bacteria are presented in table.

Table: Morphologies of some common bacteria.

	Species	Gram reaction	Cell morphology	Cell arrangement	Claim to fame
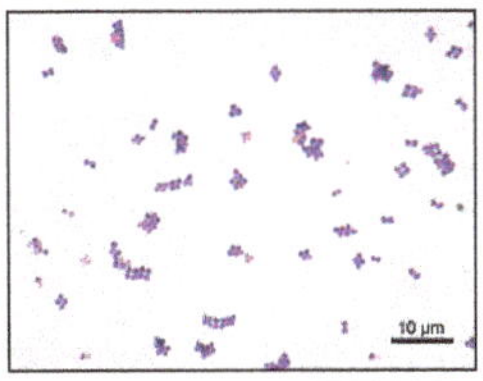	Staphylococcus aureus	Positive	Cocci	Clusters	Numerous human infections
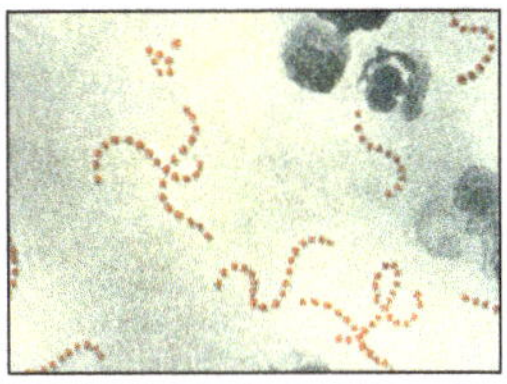	Streptococcus pyogenes	Positive	Cocci	Chains	Strep throat
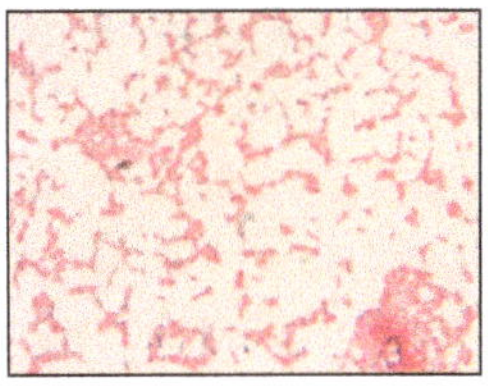	Escherichia coli	Negative	Short bacilli	Single cells, pairs	Gastrointestinal disease
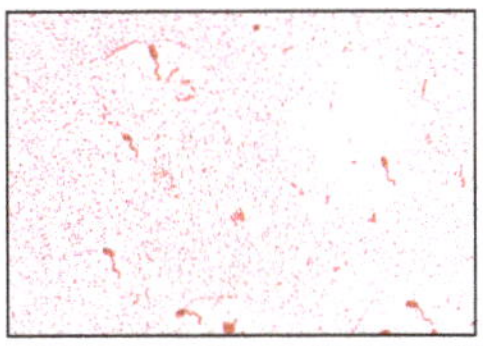	Vibrio cholerae	Negative	Curved bacilli	Single cells, pairs	Gastrointestinal disease (Cholera)
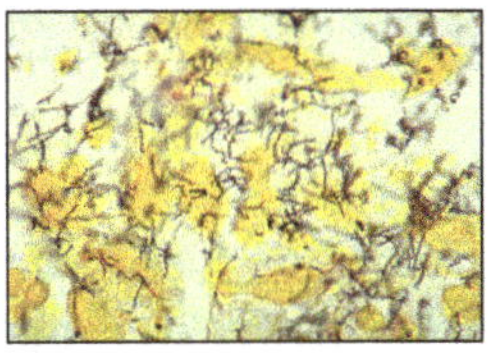	Treponema pallidum	Negative	Pirillum	Single cells	Syphilis

Arrangement of the Cell Envelopes of Gram Positive and Gram Negative Bacteria

The various layers surrounding a bacterial are collectively referred to as the cell envelope. While these include layers common to both Gram positives and Gram negatives such as the cell membrane and cell wall, there are fundamental differences that delineate the two major groups of bacteria. Figure is a diagram depicting typical Gram positive and Gram negative envelopes.

The cell membrane is a lipid bilayer containing numerous embedded proteins. The proteins, as with membrane proteins of other organisms, function in selective transport of solutes, generation of a membrane potential, ATP generation by electron transport, secretion of hydrolytic proteins, secretion of toxins and other pathogenicity factors (in some organisms), signal transduction and biosynthesis.

In Gram positive bacteria the cell membrane is surrounded by multiple layers of peptidoglycan up to 80 nm thick. Although there are many types of peptidoglycan, all are a polymer of disaccharides, one of which is always N-acetylmuramic acid, cross-linked by short peptides. In Staphylococcus aureus the peptidoglycan is a repeating unit of N-acetylglucosamine linked to N-acetylmuramic acid. The N-acetylglucosamine residues are all linked to the same 5 amino acid peptide although the peptides vary between organisms. The peptide of S. aureus is shown in figure. Peptide crossbridges are formed between the peptidoglycan chains. In S. aureus, peptide crossbridges are formed between the D-alanine of one tetrapeptide and L-lysine of the adjacent peptide chain. Despite the thickness and extensive crosslinking of the peptidoglycan layers, the cell wall is porous enough to allow diffusion of solutes to the cell membrane.

A second unique feature of the Gram positive envelope is that they often contain water soluble polymers containing ribitol or glycerol that are covalently linked to peptidoglycan (called teichoic acids) or to the cell membrane (referred to as lipoteichoic acids). The teichoic and lipoteichoic acids are major surface antigens that can be used to distinguish bacterial serotypes. The precise functions of teichoic and lipoteichoic acids are not completely understood but have been implicated in promoting bacterial adherence to mammalian cells. Other proposed functions are that teichoic acids play a role in magnesium acquisition or that they form a permeability barrier outside of the cell wall in much the same way the outer membrane functions in Gram negative organisms.

Gram negatives also have some unique envelope structures, the most obvious of which are an outer membrane overlying the cell wall and a region between the outer leaflet of the cell membrane and the inner leaflet of the outer membrane referred to as the periplasmic space or periplasm. The inner leaflet of the outer membrane closely resembles that of the cell membrane but the outer leaflet is composed of molecules of lipopolysaccharide (LPS). LPS is a complex amphipathic molecule containing a complex glycolipid called lipid A, attached to a polysaccharide tail. The lipid A portion of LPS is in the interior of the outer membrane. The polysaccharide portion of LPS is composed of a core region which is similar in all Gram negative LPS and a more distal region containing repeating units of tri-, penta- or treptasaccharides. The most distal region is known as the O antigen and varies greatly between organisms including members of the same species. The latter is the basis of the O antigen serotyping system. LPS is also known as endotoxin due to its toxicity to animal species. Magnesium ions are immobilized by adjacent LPS molecules. The magnesium is believed to provide stability to the outer membrane. The outer membrane does provide a barrier to hydrophobic and large molecules such as proteins. It also provides intrinsic resistance to potentially toxic substances such as biosalts and antibiotics.

The peptidoglycan layer is much thinner, approximately 10 nm, in Gram negatives than it is in Gram positives. The peptidoglycan layer lies between the inner and outer membranes in the periplasm. The periplasm contains numerous degradative enzymes used to break down large molecules for transport, detoxification enzymes, proteins for uptake of solutes, enzymes for the synthesis of cell walls, pili and in some organisms a secretion apparatus. The periplasm also contains an "osmotic buffer" composed of membrane-derived oligosaccharides which help stabilize the cell membrane, protecting the cell from lysis.

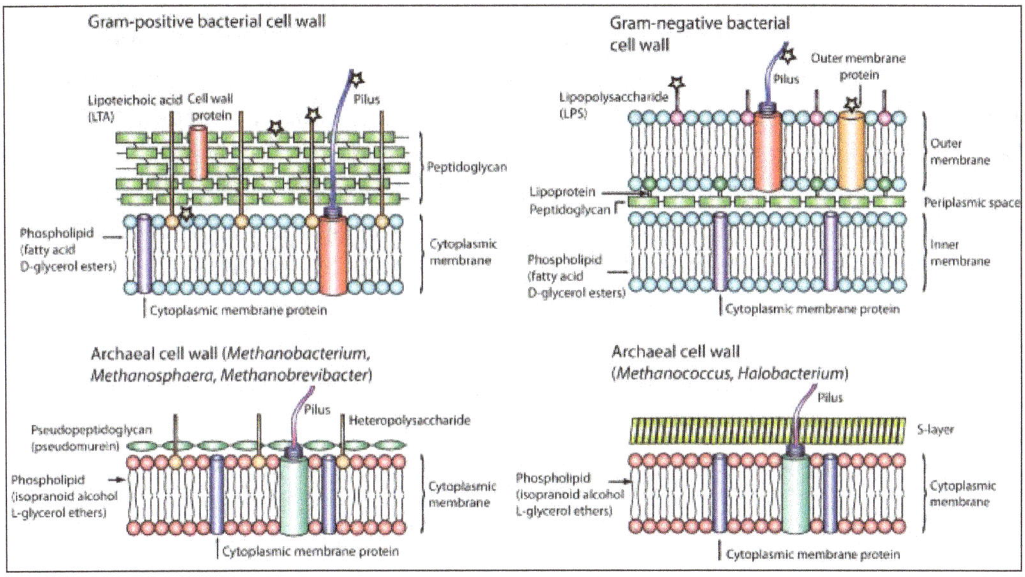

Diagrams depicting typical Gram negative (A) and Gram positive (B) cell envelopes.

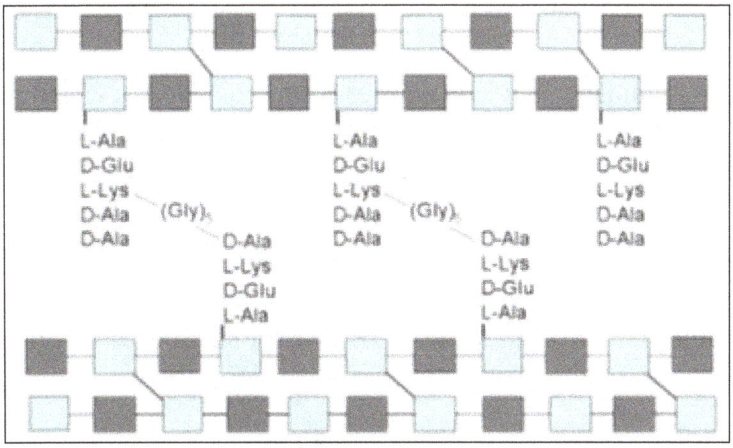

Structure of Staphylococcus aureus peptidoglycan. Peptide crosslinks between lysine and alanine residues are indicated by black lines. Blue squares indicate N-acetylglucosamine residues, gray squares indicate N-acetylmuramic acid residues.

Organisms that are neither Gram Positive or Gram Negative

A number of bacteria do not fit well in either the Gram positive or Gram negative groups. One group is the Mycobacterium species which includes M. tuberculosis the etiologic agent of tuberculosis and M. leprae which causes leprosy. These are often referred to as acid fast organisms for when stained with carbolfusion dye, their cells resist discoloration with a mixture of hydrochloric acid and alcohol. While sharing several properties with Gram positive organisms, the cell walls of mycobacteria are composed of over one-half lipid. The petpidoglycan layer is surrounded by a layer of arabinoglycan, a polysaccharide which in turn is surrounded by a layer of mycolic acid, a unique lipid. The cell envelope also contains lipoarabinomannan which is embedded in the cell membrane by phosphatidylinositol, and extends through the peptidoglycan layer. Other lipids within the cell envelope include cord factor and wax D. The outer capsule-like layer of mycobacteria contains the polysaccharides arabinomannan and mannan. The cell envelopes of mycobacteria are hydrophobic thus making the organism quite resistant to chemical agents. The hydrophobic barrier also makes uptake of nutrients difficult which might in part account for the slow growth of mycobacteria, M. tuberculosis for example has a doubling time of 18 hours under optimal growth conditions. In comparison, the doubling time of E. coli can be as low as 20 minutes.

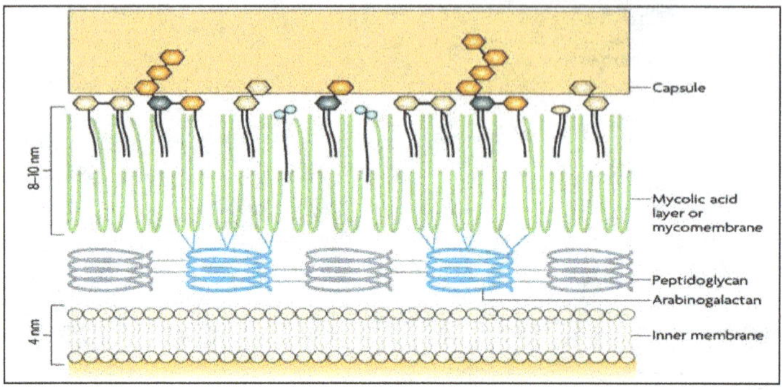

Cell envelope of mycobacteria.

Another group of organisms which is neither Gram positive nor Gram negative are the mycoplasmas. While phylogenetically related to Gram positive organisms, mycoplasmas do not synthesize peptidoglycan. The mycoplasmas are also unusual in that their cell membranes contain sterols.

It should also be noted that with some Gram positive organisms, while having the cell envelope structure of a Gram positive, will not always appear positive in a Gram stain. The Gram variable trait is often affected by growth stage of the organisms, nutrient composition, the source of the specimen, etc.

Bacterial Extracellular Polysaccharides

Many bacteria produce extracellular polymers that surround the entire cell. When these polymers are closely associated with cells they are referred to as capsules. A seemingly endless variety of chemically and immunologically distinguishable types of capsules exist. The human pathogen Streptococcus pneumoniae alone produces over one hundred different types of capsules. Most bacterial capsules are polysaccharides, but Bacillus anthracis, the causative agent of anthrax, has a polyglutamic acid (protein) capsule. Capsules have multiple functions. In pathogenic organisms capsules can inhibit phagocytosis by white blood cells, block complement activation, protect bacteria from antimicrobials, and promote adherence to host tissues. For species that produce multiple serotypes of capsules, capsules can help prevent antibodies from binding the cell. Some capsules are poor antigens, as in the case of Streptococcus pyogenes, and thus can help pathogens avoid immune surveillance.

Some bacteria produce extracellular polysaccharide polymers that are not tightly associated with cells but rather are somewhat amorphous slime layers which contain embedded bacterial cells. Such structures are referred to as biofilms. Important examples of biofilms include those formed over tooth enamel by Streptococcus mutans, which can lead to dental caries, and those formed on indwelling medical devices (e.g. catheters) by Staphylococcus epidermidis. The formation of biofilms is believed to provide pathogens protection from antibiotics and the host immune system.

Table: Extracellular polymers produced by select pathogens.

Species	Polymer	Saccharide subunits
Staphylococcus aureus	polysaccharide	Multiple types; may contain N-acetyl galactosaminuronic acid, N-acetyl-glucosaminuronic acid, N-acetyl-mannosaminuronic acid
Streptococcus pneumoniae	polysaccharide	Multiple types; may contain Glucose, galactose, rhamnose, glucuronic acid, N-acetylglucosamine
Streptococcus pyogenes	hyaluronic acid	N-acetylglucosamine, glucuronic acid
Streptococcus salivarius	levan	fructose
Streptococcus mutans	dextran (biofilm)	glucose
Escherichia coli	polysaccharide	Glucose, fucose and uronic acid
Bacillus anthracis	polypeptide	poly-D-glutamic acid
Pseudomonas aeruginosa	alginate (biofilm)	mannuronate, guluronate
Staphylococcus aureus and S. epidermidis	(biofilm)	N-acetylglucosamine

MYCOLOGY

Mycology is the study of fungi. The ability of fungi to invade plant and animal tissue was observed in early 19th century but the first documented animal infection by any fungus was made by Bassi, who in 1835 studied the muscardine disease of silkworm and proved the that the infection was caused by a fungus Beauveria bassiana.

In 1910 Raymond Sabouraud published his book *Les Teignes*, which was a comprehensive study of dermatophytic fungi. He is also regarded as father of medical mycology.

Importance of fungi: Fungi inhabit almost every niche in the environment and humans are exposed to these organisms in various fields of life.

Beneficial effects of Fungi

- Decomposition: Nutrient and carbon recycling.
- Biosynthetic factories: The fermentation property is used for the industrial production of alcohols, fats, citric, oxalic and gluconic acids.
- Important sources of antibiotics, such as Penicillin.
- Model organisms for biochemical and genetic studies. Eg: Neurospora crassa
- Saccharomyces cerviciae is extensively used in recombinant DNA technology, which includes the Hepatitis B Vaccine.
- Some fungi are edible (mushrooms).
- Yeasts provide nutritional supplements such as vitamins and cofactors.
- Penicillium is used to flavour Roquefort and Camembert cheeses.
- Ergot produced by claviceps purpurea contains medically important alkaloids that help in inducing uterine contractions, controlling bleeding and treating migraine.
- Fungi *(Leptolegnia caudate* and *Aphanomyces laevis*) are used to trap mosquito larvae in paddy fields and thus help in malaria control.

Harmful effects of Fungi

- Destruction of food, lumber, paper, and cloth.
- Animal and human diseases, including allergies.
- Toxins produced by poisonous mushrooms and within food (Mycetism and Mycotoxicosis).
- Plant diseases.
- Spoilage of agriculture produce such as vegetables and cereals in the godown.
- Damage the products such as magnetic tapes and disks, glass lenses, marble statues, bones and wax.

General Properties of Fungi

- They are eukaryotic; cells contain membrane bound cell organelles including nuclei, mitochondria, golgi apparatus, endoplasmic reticulum, lysosomes etc. They also exhibit mitosis.
- Have ergosterols in their membranes and possesses 80S ribosomes.
- Have a rigid cell wall and are therefore non-motile, a feature that separates them from animals. All fungi possess cell wall made of chitin.

- Are chemoheterotrophs (require organic compounds for both carbon and energy sources) and fungi lack chlorophyll and are therefore not autotrophic.
- Fungi are osmiotrophic; they obtain their nutrients by absorption.
- They obtain nutrients as saprophytes (live off of decaying matter) or as parasites (live off of living matter).
- All fungi require water and oxygen and there are no obligate anaerobes.
- Typically reproduce asexually and/or sexually by producing spores.
- They grow either reproductively by budding or non-reproductively by hyphal tip elongation.
- Food storage is generally in the form of lipids and glycogen.

Classification of Fungi

Fungi were initially classified with plants and were a subject of interest for botanists; hence the influence of botany can be seen on their classification. In 1969 R.H Whittaker classified all living organisms into five kingdoms namely Monera, Protista, Fungi, Plantae and Animalia.

Traditionally the classification proceeds in this fashion:

Kingdom - Subkingdom - Phyla/phylum - Subphyla - Class - Order - Family - Genus- Species This classification is too complicated to be dealt here.

There are alternate and more practical approaches, one based on sexual reproduction and the other based on morphology of the thallus (vegetative structure).

Based on Sexual Reproduction

- Zygomycetes: Which produce through production of zygospores.
- Ascomycetes: Which produce endogenous spores called ascospores in cells called asci.
- Basidiomycetes: Which produce exogenous spores called basidiospores in cells called basidia.
- Deuteromycetes (Fungi imperfecti): Fungi that are not known to produce any sexual spores (ascospores or basidiospores): This is a heterogeneous group of fungi where no sexual reproduction has yet been demonstrated.

Based on Morphology

- Moulds (Molds): Filamentous fungi Eg: *Aspergillus sps*, *Trichophyton rubrum.*

- Yeasts: Single celled cells that buds Eg: *Cryptococcus neoformans, Saccharomyces cerviciae.*
- Yeast like: Similar to yeasts but produce pseudohyphae Eg: *Candida albicans.*

Dimorphic: Fungi existing in two different morphological forms at two different environmental conditions. They exist as yeasts in tissue and in vitro at 37^0C and as moulds in their natural habitat and in vitro at room temperature. Eg: Histoplasma capsulatum, Blastomyces dermatidis, Paracoccidiodes brasiliensis, Coccidioides immitis.

Some 200 "human pathogens" have been recognized from among an estimated 1.5 million species of fungi.

Morphology of Fungi

Fungi exist in two fundamental forms; the filamentous (hyphal) and single celled budding forms (yeast). But, for the classification sake they are studied as moulds, yeasts, yeast like and dimorphic fungi.

All fungi have typical eukaryotic morphology. They have rigid cell wall composed of chitin, which may be layered with mannans, glucans and other polysaccharides in association with polypeptides. Some lower fungi possess cellulose in their cell wall. Some fungi such as Cryptococcus and yeast form of *Histoplasma capsulatum* possess polysaccharide capsules that help them to evade phagocytosis.

Inner to the cell wall is the plasma membrane that is a typical bi-layered membrane in addition to the presence of sterols. Fungal membranes possess ergosterol in contrast to cholesterol found in mammalian cells. The cytoplasm consists of various organelles such as mitochondria, golgi apparatus, ribosomes, endoplasmic reticulum, lysosomes, microtubules and a membrane enclosed nucleus. A unique property of nuclear membrane is that it persists throughout the metaphase of mitosis unlike in plant and animal cells where it dissolves and re-forms. The nucleus possesses paired chromosomes.

Moulds

The thallus of mould is made of hyphae, which are cylindrical tube like structures that elongates by growth at tips. A mass of hyphae is known as mycelium. It is the hypha that is responsible for the filamentous nature of mould. The hyphae may be branched or unbranched. They may be septate or aseptate. Hyphae usually have cross walls that divide them into numerous cells. These cross walls, called septa have small pores through which cytoplasm is continuous throughout the hyphae. Therefore all hyphal fungi tend to be coenocytic (multinucleate). With exception of zygomycetes (Rhizopus, Mucor), all moulds are septate. Non-septate hyphae are considered to be more primitive because if a hyphal strand is damaged the entire strand dies. When a septate hyphal strand is damaged, the pores between adjacent compartments can be plugged, thus preventing death of the whole hyphal strand.

Mycelium are of three kinds:

- Vegetative mycelium are those that penetrates the surface of the medium and absorbs nutrients.
- Aerial mycelium are those that grow above the agar surface.
- Fertile mycelium are aerial hyphae that bear reproductive structures such as conidia or sporangia.

Since hypha is the structural unit of mould, the mycelium imparts colour, texture and topography to the colony. Those fungi that possess melanin pigments in their cell wall are called phaeoid or dematiaceous and their colonies are coloured grey, black or olive. Examples are species of Bipolaris, Cladosporium, Exophiala, Fonsecaea, Phialophora and Wangiella Those hyphae that don't possess any pigment in their cell wall are called hyaline. Hyphae may have some specialised structure or appearance that aid in identification.

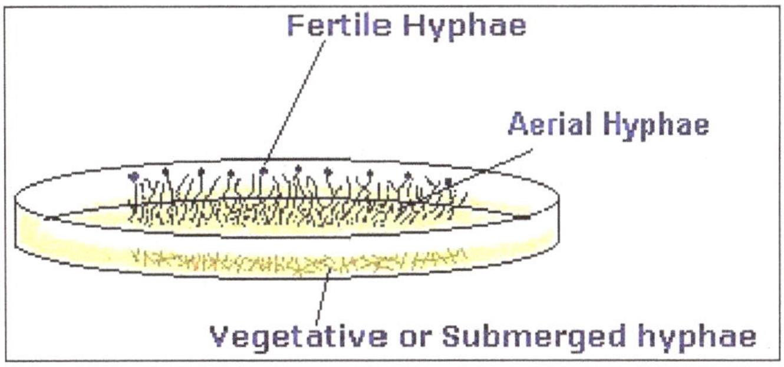

Some of these are:

- Spiral hyphae: These are spirally coiled hyphae commonly seen in *Trichophyton mentagrophytes*.
- Pectinate body: These are short, unilateral projections from the hyphae that resemble a broken comb. Commonly seen in *Microsporum audouinii*.
- Favic chandelier: These are the group of hyphal tips that collectively resemble a chandelier or the antlers of the deer (antler hyphae). They occur in *Trichophyton schoenleinii* and *Trichophyton violaceum*.
- Nodular organ: This is an enlargement in the mycelium that consists of closely twisted hyphae. Often seen in *Trichophyton mentagrophytes* and *Microsporum canis*.
- Racquet hyphae: There is regular enlargement of one end of each segment with the opposing end remaining thin. Seen in *Epidermophyton floccosum, Trichophyton mentagrophytes*.
- Rhizoides: These are the root like structures seen in portions of vegetative hyphae in some members of zygomycetes.

- There are structures in the hyphae, which arise out of modification of a single cell and transform into thick walled resting cells. Chlamydospore (or chlamydoconidia), which are produced by *Trichophyton schoenleinii* and *Trichophyton verrucosum* are thick walled cells that are larger than other cells and arranged singly or in groups. In some fungi such as *Trichosporon beigeilli* and *Coccidioides immitis* some alternating cells become thick walled and subsequently the intervening cells disintegrate leaving behind arthrospores (or arthroconidia).

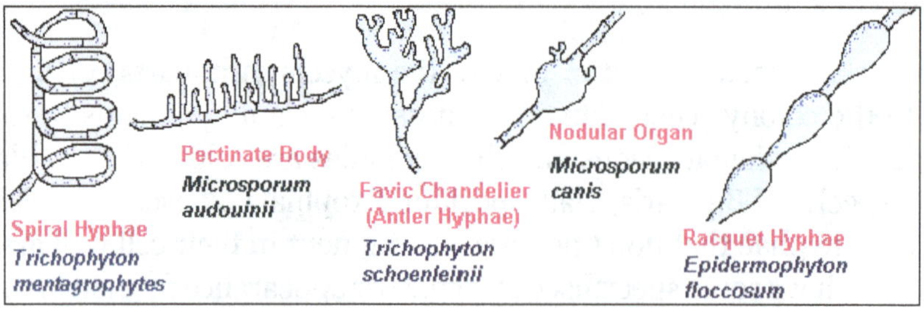

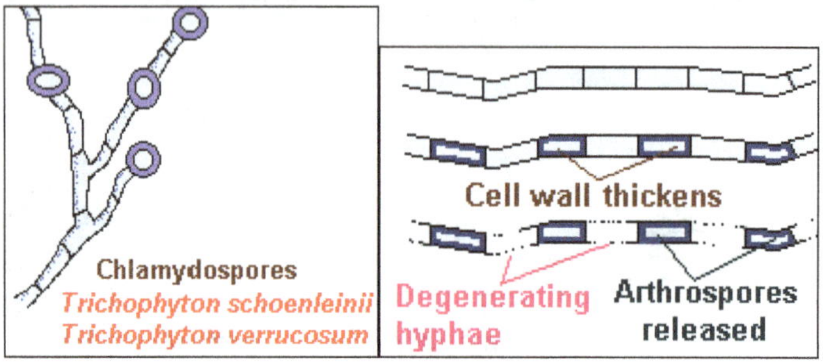

Yeasts

Yeasts are unicellular spherical to elipsoid cells. They eproduce by budding, which result in blastospore (blastoconidia) formation. In some cases, as the cells buds the buds fail to detach and elongate thus forming a chain of elongated hyphae like filament called pseudohyphae. This property is seen in *Candia albicans*. The same species also have the ability to produce true hypha, which is seen as germ tube. The difference between the two is that there is a constriction in psueudohyphae at the point of budding, while the germ tube has no constriction.

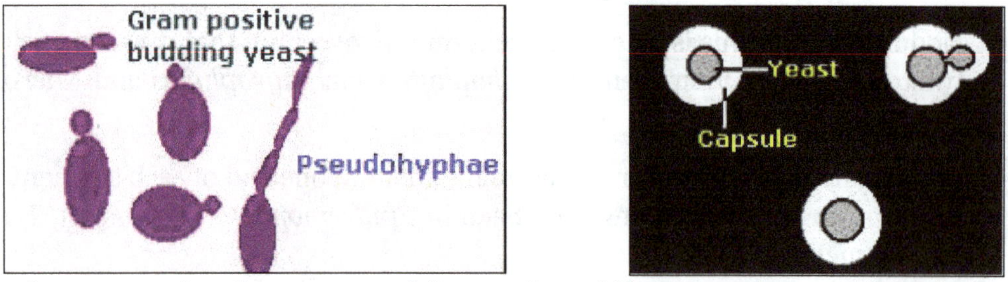

Some yeast such as Cryptococcus and the yeast form of *Blastomyces dermatatidis* produce polysaccharide capsule. Capsules can be demonstrated by negative staining methods using India ink or Nigrosin. The capsule itself can be stained by Meyer Mucicarmine stain.

Some yeasts are pigmented. *Rhodotorula sps* produces pink colonies due to carotenoid pigments while some yeasts such as *Phaeoannellomyces werneckii* and *Piedraia hortae* are dematiaceous, producing brown to olivaceous colonies.

True yeasts such as *Saccharomyces cerviciae* don't produce pseudohyphae. Yeast-like fungi may be basidiomycetes, such as *Cryptococcus neoformans* or ascomycetes such as *Candida albicans*.

Reproduction in Fungi

Fungi reproduce by asexual, sexual and parasexual means.

Asexual reproduction is the commonest mode in most fungi with fungi participating in sexual mode only under certain circumstances. The form of fungus undergoing asexual reproduction is known as anamorph (or imperfect stage) and when the same fungus is undergoing sexual reproduction, the form is said to be teleomorph (or perfect stage). The whole fungus, including both the forms is referred as holomorph. *(Taxonomically, the teleomorph or the holomorph is used, but practically it is more convenient to use the anamorph).*

Asexual Reproduction

Asexual propagules are termed either spores or conidia depending on their mode of production. Asexual spores are produced following mitosis where as sexual spores are produced following meiosis.

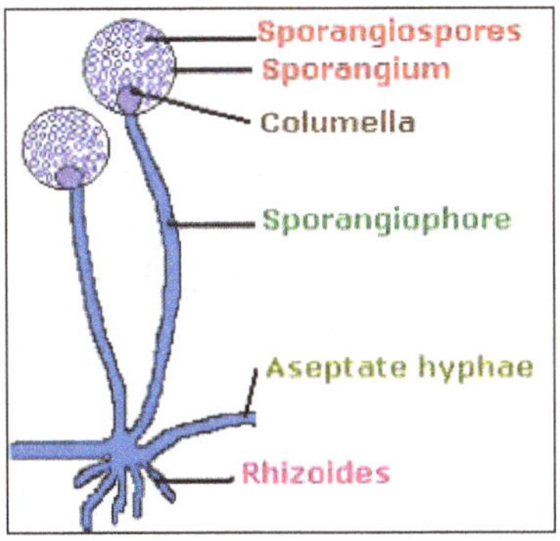

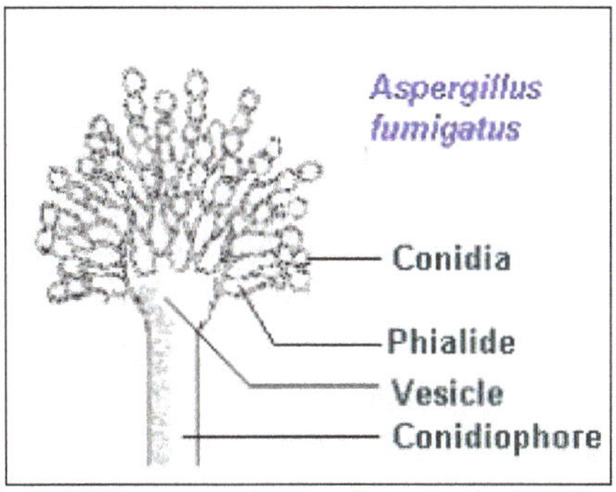

The asexual spores of zygomycetes, which are known as sporangiospores form within sac like structure known as sporangia. The sporangiospores result from the mitotic cleavage of cytoplasm in the sporangium. The sporangia are borne on special hyphae called sporangiophore. This endogenous process of spore formation within a sac is known as sporogenesis.

Conidia arise either by budding off conidiogenous hyphae or by differentiation of preformed hyphae. These develop following mitosis of a parent nucleus and are formed in any manner except involving cytoplasmic cleavage. This exogenous process is known as conidiogenesis, a process that occurs both in yeasts and moulds. Conidia are borne on specialised structures called conidiophore.

Conidia production may be blastic or thallic. In blastic development the conidium begins to enlarge and a septum is formed. Here the conidium originates from part of parent. In thallic mode of development the conidium is differentiated by a septum before its differentiation. Thus the conidium results from the conversion of entire parent cell into the conidium.

The cell that gives rise to a conidium is called a conidiogenous cell. Conidiophores are specialised hyphae that bear conidia or conidiogenous cells. In many cases conidiogenous cells are referred as phialides.

Sexual Reproduction

Sexual propagules are produced by the fusion of two nuclei that then generally undergo meiosis.

The first step in sexual methods of reproduction involves plasmogamy (cytoplasmic fusion of two cells). The second step is karyogamy (fusion of two compatible nuclei), resulting in production of diploid or zygote nucleus. This is followed by genetic recombination and meiosis. The resulting four haploid spores are said to be sexual spores, e.g. zygospores, ascospores and basidiospores.

If a sexual spore is produced only by fusion of a nucleus of one mating type with a nucleus of another mating type (+ and - strains), the fungus is said to be heterothallic. In contrast, homothallic moulds produce sexual spores following the fusion of two nuclei from the same strain. For sexual reproduction to occur, two compatible isolates are required.

Zygospores, which are the sexual spores of zygomycetes are round, thick walled reproductive structures that result from the union of two gamentagia. Ascomycetes produce sexual spores called ascospores in a special sac like cell known as ascus. In basidiomycetes the basidiospores are released from basidium, which is the terminal cell of a hyphae.

Parasexual Reproduction

Parasexual reproduction, first seen in Aspergillus is known to occur in basidiomycetes, ascomycetes and deuteromycetes. The process involves genetic recombination without the requirement of specific sexual structures.

Importance of Spores

Biological

- Allows for dissemination.
- Allows for reproduction.

- Allows the fungus to move to new food source.
- Allows fungus to survive periods of adversity.
- Means of introducing new genetic combinations into a population.

Practical

- Rapid identification (also helps with classification).
- Source of inocula for human infection.
- Source of inocula for contamination.

Zygomycetes

Commonly known as bread moulds, these are fast growing, terrestrial, largely saprophytic fungi. Hyphae are coenocytic and mostly aseptate. Asexual spores include chlamydoconidia, conidia and sporangiospores. Sporangiophores may be simple or branched. Sexual reproduction involves producing a thick-walled sexual resting spore called a zygospore.

- Medically important orders and genera include:
- Entomophthorales: Conidiobolus and Basidiobolus are involved in subcutaneous zygomycosis
- Mucorales: Rhizopus, Mucor, Rhizomucor, Absidia and Cunninghamella are involved in subcutaneous and systemic zygomycosis (formerly called Mucormycosis).

Basidiomycetes

They exist as saprobes and parasites of plants. Hyphae are dikaryotic and can often be distinguished by the presence of clamp connections over the septa. Sexual reproduction is by the formation of exogenous basidiospores, typically four, on a basidium. Occasional species produce conidia but most are sterile.

Genera of medical importance include:

- Teleomorph of *Cryptococcus neoformans*, which is *Filobasidiella neoformans*.
- Agents of basidiomycosis such as Coprinus and Schizophyllium.
- Mushroom poisoning by Aminita, Lepiota, Coprinus and Psilocybe etc.

Ascomycetes

They exist as saprophytes and parasites of plants. Hyphae are septate with simple septal pores. Asexual reproduction is by conidia. Sexual reproduction is by the formation of endogenous ascospores, typically eight, in an ascus.

Medically important genera include the:

- Teleomorphs of known pathogenic fungi e.g. Arthroderma (of Trichophyton and Microsporum), *Ajellomyces dermatitidis* (of *Blastomyces dermatitidis*), *Pseudallescheria boydii* (of *Scedosporium apiospermum*).
- Agents of mycetoma, like Leptosphaeria.
- Agents of black piedra, like *Piedraia hortae*.

Deuteromycetes

Deuteromycetes are also known as Fungi Imperfecti because of absence of sexually reproducing forms (teleomorph or perfect stage). As their teleomorph continue to be discovered, they would be classified among the previous categories, until then this remains an artificial and heterogeneous group.

There are three classes of Fungi Imperfecti.

- Blastomycetes: These include asexual budding forms of Cryptococcus, Candida, Torulopsis and Rhodotorula. Depending on the presence of melanin in their cell walls, they may be non-dematiaceous or dematiaceous.
- Hyphomycetes: A class of mycelial moulds which reproduce asexually by conidia on hyphae. Hyphae are septate. This class contains the majority of medically important fungi. Dematiaceous hyphomycetes are those conidial fungi that produce dark brown, green-black, or black colonies and are the causative agents of phaeohyphomycosis. Hyaline hyphomycetes include those conidial fungi, which are not darkly pigmented; colonies may be colourless or brightly coloured. These include the agents of hyalohyphomycosis, aspergillosis, dermatophytosis and the dimorphic pathogens, like *Histoplasma capsulatum*.
- Coelomycetes: These produce acervuli, which are tightly bound mats of hyphae on which conidia are produced.

Pathogenesis of Fungal Diseases (Mycoses)

Most fungi are saprophytic or parasitic to plants and are adapted to their natural environment. Infection in humans is a chance event, occurring only when conditions are favourable. Except for few fungi such as the dimorphic fungi that cause systemic mycoses and dermatophytes, which are primary pathogens, the rest are only opportunistic pathogens.

Human body is a hostile environment and offers great resistance to fungal invasion. Most fungi are saprophytic and their enzymatic pathways function more efficiently at the redox potential of non-living substrates than at the relatively more reduced state of living metabolizing tissue. Some fungi such as Candida and Malasezzia have adapted to human environment and exist as commensals.

The complex interplay between fungal virulence factors and host defence factors will determine if a fungal infection will cause a disease. Infection depends on inoculum size and the general immunity of the host.

Fungal Pathogenicity (Virulence Factors)

- Ability to adhere to host cells by way of cell wall glycoproteins.
- Production capsules allowing them to resist phagocytosis.
- Production of a cytokine called GM-CSF by Candida albicans that suppress the production of complement.
- Ability to acquire iron from red blood cells as in Candida albicans.
- Ability to damage host by secreting enzymes such as keratinase, elastase, collagenase.
- Ability to resist killing by phagocytes as in dimorphic fungi.
- Ability to secrete mycotoxins.
- Having a unique enzymatic capacity.
- Exhibiting thermal dimorphism.
- Ability to block the cell-mediated immune defences of the host.
- Surface hydrophobicity.

Host Defence Factors

- Physical barriers, such as skin and mucus membranes.
- The fatty acid content of the skin.
- The pH of the skin, mucosal surfaces and body fluids.
- Epithelial cell turnover.
- Normal flora.
- Chemical barriers, such as secretions, serum factors.
- Most fungi are mesophilic and cannot grow at $37^{O}C$.
- Natural Effector Cells (polymorphonuclear leucocytes) and the Professional Phagocytes (monocytes and macrophages).

Factors Predisposing to Fungal Infections

- Prolonged antibiotic therapy.

- Underlying disease (HIV infection, cancer, diabetes, etc.)
- Age.
- Surgical procedures.
- Immunosuppressive drugs.
- Irradiation therapy.
- Indwelling catheters.
- Obesity.
- Drug addiction.
- Transplants.
- Occupation.

Immunity to Fungal Infections

Mechanism of immunity to fungal infections can be innate or acquired. The non-specific immunity includes the physical barriers offered by skin and mucus membranes along with their secretions and normal flora. The pH, body temperature and serum factors along with phagocytic cells play an important part in providing non-specific immunity. Even though body mounts both humoral and cell mediated immunity, it is the latter that is the mainstay of host defence.

Cell Mediated Immunity

Immunity is provided non-specifically be effector cells (polymorphonuclear leucocytes) and professional phagocytes (monocytes and macrophages) and specifically by T lymphocytes. The phagocytes are very important in defence against Candia, Aspergillus and Zygomycetes as is evidenced by their severity in granulomatous diseases, myeloperoxidase deficiency and cytotoxic chemotherapy.

Expression of T-cell-mediated immunity to fungi includes:

- Delayed-type hypersensitivity.
- Contact allergy.
- Chronic granulomatous reactions.

Humoral Immunity

Even though antibodies are produced against many fungi, their role in protection is not very clear. However, antibodies help in clearing fungal pathogens through opsonisation, which is important against Candida and Cryptococcus. Another component of humoral immunity is the complement, which can act as opsonins and may even cause damage to their cells through complement activation. Antibodies are important to fungal sero-diagnosis.

Hypersensitivity

As a result of dermatophyte infection some fungus-free skin lesions of variable morphology occur elsewhere on the body, which are thought to result from hypersensitivity to the fungus. These reactions are called "id reaction". These reactions are also seen n Candida infections. An inflamed boggy lesion of the scalp called the kerion may result from a strong immune reaction to the dermatophyte. Granulomas due to intracellular fungi represent delayed hypersensivities. Many fungi are significant allergens to humans, the allergens being spores, conidia, hyphae and other fungal products. On inhalation they may produce allergic pulmonary diseases such as allergic bronchopulmonary aspergillosis, farmer's lung, maple bark stripper's lung, bronchial asthma etc, which may be Type I or III hypersensitivity.

Fungal Diseases

Mycoses can be conveniently studied as:

Superficial mycoses:

- Superficial phaeohyphomycosis.
- Tinea versicolor.
- Black piedra.
- White piedra.

Cutaneous mycoses:

- Dermatophytosis.
- Dermatomycosis.

Subcutaneous mycoses:

- Chromoblastomycosis.
- Rhinosporidiasis.
- Mycetoma.
- Sporotrichosis.
- Subcutaneous phaeohyphomycosis.

- Lobomycosis.

Systemic (deep) mycoses:

- Blastomycosis.
- Histoplasmosis.
- Coccidioidomycosis.
- Paracoccidioidomycosis.

Opportunistic mycoses:

- Candidiasis.
- Cryptococcosis.
- Aspergillosis.

Other mycoses:

- Otomycosis.
- Occulomycosis.
- Fungal allergies.
- Mycetism and mycotoxicosis.

Laboratory Diagnosis of Mycoses

Specimen collection: specimen collection depends on the site affected. Different specimens include hair, skin scrapings, nail clippings, sputum, blood, CSF, urine, corneal scraping, discharge or pus from lesions and biopsy.

- All specimens must be transported to the laboratory without any delay to prevent bacterial overgrowth. In case of delay specimens except skin specimen, blood and CSF may be refrigerated for a short period.
- Infected hairs may be plucked using forceps. Those hairs that fluoresce under Wood's lamp may be selectively plucked. Hairs may be collected in sterilized paper envelopes.
- Surface of the skin must be disinfected with spirit before specimen collection. The advancing edge of the lesion is scraped with the help of a blunt forceps and collected in sterilized paper envelopes.
- Discoloured or hyperkeratotic areas of nail may be scraped or diseased nail clipping may be collected in sterilized paper envelopes.

- Specimens from mucus membranes (oral) must be collected by gentle scraping and transported to laboratory in sterile tube containing saline. Swabs may be collected from vagina.
- Corneal scrapings may be collected using a fine needle and inoculated at bed-side.
- Pus may be collected by aspiration; use of cotton swabs may give false positive microscopic results.
- Clean catch urine may be collected in a sterile wide-mouthed container.
- Biopsy specimens must be transported in saline.

In certain cases, pus or exudates must be looked for presence of granules.

Microscopy: Microscopy is used to observe clinical specimens for the presence of fungal elements or to identify the fungus following culture. In the latter case, lactophenol cotton blue is stain of choice, which stains the fungal elements blue. Direct examination of clinical specimens could be stained or unstained.

- Wet mount: Candida may be observed in urine wet mounts.
- 10-20% KOH mount: Several specimens are subjected to KOH mount for direct examination. The material is mixed with 20% KOH on a slide and a cover slip is placed. The slide is then gently heated by passing through the flame 2-3 times. The slide is observed on cooling. KOH serves to digest the protein debris and clears keratinised tissue and increases the visibility. Addition of Dimethyl sulphoxide (DMSO) permits rapid clearing in the absence of heat.
- Calcofluor white: This is a fluorescent dye, which binds selectively to chitin of the fungal cell wall. The specimen then can be observed under fluorescent microscope.
- India Ink: Capsules of Cryptococcus neoformans can be demonstrated by this negative staining technique.
- Periodic Acid-Schiff (PAS) stain: On staining by this stain, fungal elements appear bright magenta coloured while the background stains green. It is useful in staining tissue specimens.
- Giemsa's stain: It is particularly useful in the detection of *Histoplamsa capsulatum* in the bone marrow smears.
- Haematoxylin and Eosin (H&E) stain: Useful for staining tissue sections.
- Gomori's methenamine silver nitrate (GMS) stain: Outlines of the fungi are black, internal parts stain pink- black while the background stains light green.

Candida and Aspergillus may be missed in H&E stained sections, therefore GMS stained sections are essential for tissue pathology.

- Gridley's stain: It stains hyphae and yeasts dark blue-pink, tissues deep blue and background yellow.
- Meyer mucicarmine stain: Capsules of *C. neoformans* and inner walls of *Rhinosporidium seeberi*'s sporangium are stained pink.
- Gram stain: Candida is best demonstrated in clinical specimen by Gram stain.
- Masson-Fontana stain is helpful in staining phaeoid (dematiaceous) fungi in tissue.
- Immunofluorescence: Monoclonal antibody labelled with fluorescent dyes can be used to detect several fungi in the clinical specimens.

Culture: One of the most common media used to culture fungi in laboratory is Sabouraud's Dextrose Agar (SDA). It consists of peptone, dextrose and agar. High concentration of sugar and a low pH (4.5-5.5) prevents growth of most bacteria and makes it selective for fungi. Emmon's modification of SDA contains 2% dextrose and has pH of 6.8.

Other basal media to grow fungi include Potato Dextrose Agar, Malt Extract Agar etc. Most fungi are able to grow at room temperature while few pathogenic fungi (e.g, Cryptococcus, dimorphic fungi) can grow at $37^{O}C$. Saprophytic fungi grow much quickly than pathogenic fungi (e.g, dermatophytes). In such situations the saprophytic fungi can be inhibited by the addition of cycloheximide (actidione) to the SDA. Addition of antibi-otics such as Chloramphenicol, Gentamicin or Streptomycin to SDA serves to inhibit bacterial multiplication. An example of SDA with cycloheximide and Chloramphenicol is Mycosel agar.

Other specialized media used for different fungi include:

- Brain Heart Infusion Agar general isolation of fungi and conversion of dimorphic fungi.
- Inhibitory Mould Agar, an isolation medium with Chloramphenicol to suppress most bacteria.
- Caffeic Acid Agar and Birdseed Agar for isolation of Cryptococcus neoformans.
- Corn Meal Agar: Enhances production of chlamydospores in Candida albicans and formation of conidia in fungi.
- Trichophyton Agars: Used for selective identification of Trichophyton species.

- Dermatophyte Test Medium: Used for recovery of dermatophytes from clinical specimens.
- Sabhi Medium: Isolation of Histoplasma capsulatum.
- 'CHROM agar Candida' is useful in identification of Candida species.

Conversion of mould to yeast phase must be demonstrated in vitro for identification of dimorphic fungi. Since some fungi grow slowly cultures should not be discarded for 4-6 weeks. Fungi are identified on the basis of colony morphology (including pigmentation) and microscopic observation by tease-mount preparation or slide culture technique.

Serology: Detection of anti-fungal antibody is helpful in diagnosis of sub-cutaneous and systemic mycoses, prognosis and response to anti-fungal drugs. Different serologic techniques that are used include agglutination, immunodiffusion, counter-immuno-electrophoresis, complement fixation test, immunofluorescence, RIA and ELISA.

Antigen detection: It is particularly useful in the diagnosis of cryptococcal meningitis from CSF specimens. The test is performed by Latex Agglutination or immunodiffusion tests. It is also helpful in the detection of Aspergillus and Candida antigens in systemic infections.

Skin tests: Delayed hypersensitivity reactions to fungal antigens can be demonstrated by skin tests. A positive skin does not necessarily indicate an active infection; it only indicates sensitization of the individual. Hence, its value is in epidemiological studies than diagnosis. These tests may be performed in Histoplasmosis, Candidiasis, Sporotrichosis, Coccidioidomycosis, Blastomycosis, Paracoccidiodomycosis and dermatophytosis.

Molecular techniques: Newer techniques such as DNA hybridization, PCR are useful in diagnosis of mycoses in a shorter period as well as detect those fungi that are difficult or dangerous to cultivate in vitro.

PROTOZOOLOGY

The science of the study of the protozoa is called Protozoology. Protozoans are the first group of organized but simplest animals in the animal kingdom. Modern Zoologists thought that the higher animals have been evolved from the single celled protozoans through the evolutionary processes.

Hence protozoans are the first group of animals that appear on this earth surface. So they are regarded as the most primitive animals in the animal kingdom. They are unicellular without any tissue grade of organization but multiply asexually by binary or multiple fission or by budding, in some cases sexual reproduction by conjugation.

Morphology of Protozoology

They are very minute and microscopic animals. The structure of a protozoan "cell" consists of cytoplasm body and nucleus. The cytoplasm is differentiated into two parts—the outer hyaline portion, ectoplasm; function is locomotive, protective and sensory and the inner granular portion, the endoplasm, function is nutritive and reproductive.

Generally the protozoans possess definite locomotory organs, such as flagella, cilia or pseudopodia. In some of them cysts are formed during extreme condition. In protozoa free living, parasitic or symbiotic nature of animals are noticed. Contractile vacuole is present in many protozoans. Nutrition in protozoans generally takes place by holozoic process, but in some cases saprozoic or holophytic, even parasitic mode of nutrition also occur.

Life Cycle of Protozoology

Protozoan can multiply by asexual method of reproduction for a long time. They may switch on to sexual method or undergoes cystic change to its host. In parasitic protozoa both asexual and sexual reproductions occur in a different host. This process is known as alternation of generation.

Amoebae, intestinal flagellates, Balantidium coli, etc. are monogenetic i.e. require one host to complete their life cycles. The parasites may adopt in a different host as well and transfer themselves passively in cystic stage. Again some parasitic protozoa like Leishmania, Trypanosoma and Plasmodia, etc. are digenetic i.e. require two hosts to complete their life cycles.

PARASITOLOGY

Parasitology is the study of species that depend on others for survival that cause harm to the species on which it depends. They can be plant or animal. The former is a type of plant that depends on another plant for its life cycle. Those in the animal kingdom are broadly broken down into three groups:

- Protozoa, the animal-like single-cell organisms that can be ingested or transmitted by insect bites. Malaria is an example of this. Mosquitoes carry the protozoa which then infects the host when the mosquito feeds, making its way into the bloodstream.

- Parasitic worms called helminths such as the tapeworm which lives inside the human gut, typically ingested from eating undercooked meat from an infected animal. Some parasitic worms in tropical countries can burrow into the skin of the foot.

- Arthropods such as insects and arachnids that carry disease or through their biological functions, cause them. The most common example in the west today are ticks and flea - human and animal. This group also includes the sexually-transmitted pubic lice.

Parasitic plants have roots that have evolved to penetrate the roots or stems of another plant and leech from its nutrients. Example of parasitic plants includes mistletoe, the Australian Christmas Tree, and Yellow Rattle.

As most parasites are considerably smaller than the species on which they depend, parasitology is a subdiscipline of microbiology which is, in turn, is a subdiscipline of biology. Today, it is a multidisciplinary approach with aspects of epidemiology, genetics and genomics ecology, botany and zoology. Globally, human infection from parasites number in the billions and can be from the near-harmless (taking only nutrients and energy) to killers responsible for the deaths of millions annually. The same applies to plant biodiversity. Some are harmless to plants while others have the ability to destroy an entire season's harvest of agricultural crops. WHO estimates that around 30% of people in the world today are infected with Ascaris lumbricoides, a nematode. Parasitic diseases such as malaria have only increased despite efforts to combat them. Due to the arrival of AIDS, some have seen a resurgence such as cryptosporidiosis.

Sub-disciplines of Parasitology

As parasites are so prevalent, it was inevitable that it would break down into subdisciplines to examine different aspects of parasites and their life cycles.

Immuno-parasitology: The study of the immunology and the interactions of the immune system between a host and an invading parasite. It's largely used in the development of vaccines. Researchers look at how the parasite gets around the immune system by use of genetics, specific proteins or chemical reactions of parasites. It's been particularly successful in the last decade with many vaccines successfully developed in veterinary sciences and human health. However, investment is required for continued development in some areas. As discussed above, it is estimated that parasites represent some 50% of all biological life. Not all of these affect humans, our pets, or livestock, but improvements are expected in these areas. There is still much that researchers do not understand and there remains a challenge in examining genetic aspects of the relationship.

Medical Parasitology: This is the branch of parasitology concerned with parasites that affect humans and the medical issues that they cause. Some parasites do not cause disease; they simply leech nutrients from the infected person which cause malnutrition and weight loss. However, some carry virulent and deadly diseases. For example, the ectoparasite called "tick" can carry Lyme Disease - the most prevalent parasite-spread disease in the northern hemisphere. Mosquitoes are not parasites, but they can spread parasitical diseases such as malaria. Medical parasitology is also concerned with the development of medical treatments to hinder or neutralize the disease as well as the study of disease genetics, disease diagnosis, prevention, control, and treatment.

Paleoparasitology: This is the intersection between paleontology and parasitology. Simply, it's the study of parasites in the past, their relationships with host species, and disease vectors, epidemiology, pathology and health, paleoecology and paleoclimatology. As paleontology applies to animal species in the past, paleoparasitology is concerned with parasites of extinct animals or extinct parasites and their interactions.

Archaeoparasitology: There is some debate about whether this is a true subdiscipline of parasitology or whether it should be considered in with paleoparasitology. The "archaeo" gives it an archaeological context, meaning that this area of study concerns the interactions between humans of the past and human ancestor species, and their relationships and ecology with parasites.

Parasitic Conservation Biology: It may seem inconceivable to be concerned about the conservation status and continued existence of parasites, but they have a vital role to play in local and planetary ecology. However, they represent an enormous area of biology and their existence is vital to ecological balance on local and global systems. Those that are responsible for diseases have uses in population control to ensure species do not become imbalanced, leading to ecological damage. Due to the diseases they can cause, some governments have opted for a systemic program of eradication of the parasite itself instead of releasing funding for treatment, prevention, and cure of the conditions for which they are sometimes responsible in people, livestock, and food crops. Conservation biologists attempt to challenge the view of parasites of being (at best) useless and (at worst) a menace that would lead to better health and ecology without them. Conservationists are presently pushing for funding and awareness for greater research and appreciation of this broad area of biological research.

Parasitic Molecular Biology: Molecular biology is a cross disciplinary area of chemistry and biology, with a particular focus on biochemistry, genetics, and biology at a micro level. It examines the interactions between a cell's systems such as RNA and DNA, and the processes of synthesizing proteins with a particular focus on how these interactions take place and their regulation. Everything that applies to regular molecular biology applies here but to parasites and their interactions in their own bodies and in the host-parasite relationship. Many of these interactions have protein and DNA markers, particularly in how the parasite fools the genetic code of the host and the potential problems for the immune system. This is sometimes also known as molecular parasitology.

Parasite Ecology: Ecology is the relationship between living things in an ecosystem. We sometimes perceive it as conservationism; that plays a small part of the ecology, but only one part in a wide web of areas of research. Essentially, ecology is the study of the network of life. In parasite ecology, researchers examine the range of attributes that parasites bring to a biome and the consequences (positive and negative) or their existence. Like all other biological life, parasites have adapted and continue to adapt to changes to their environment including environmental change, immune resistance and other defensive strategies of hosts, their abilities to jump species (where possible) and the relationship between carriers who are not affected by the parasite and the hosts who are affected. Their complex lifecycle within an environment is also of interest, as is the effects they have on each species within a biome. Finally, researchers examine abundance, prevalence, and the conditions that would lead to change.

Structural Parasitology: To understand parasites, it's necessary to understand many aspects of their anatomy. Parasite molecular biology is one of those areas. Another is structural parasitology which is the study of parasite proteins through the application of technology such as NMR and x-ray crystallography. The aim is to understand 3D modeling of proteins and the molecules in a parasitical relationship - applying to the parasite, its host, and their interactions. One of the major aims here is to examine functional pathways in organisms, comparing them to their hosts. Ultimately, it is hoped that discovering and understanding these links will lead to better vaccines for prevention, treatment of disease, and removal of the symptoms and diseases that they sometimes cause. In particular, it is hoped that the control and treatment of malaria will most benefit from this research area.

Quantitative Parasitology: Quantitative parasite ecology, the term "qualitative" refers to the use of applied mathematics and data modelling, usually involving statistics and distribution, in the examination of something. That is no different for parasitology. Each parasite species will tend to have an aggregated distribution among hosts. What this means is that most potential hosts carry few or no parasites while a relatively small number carry a large number of individual parasites. The bell curve is heavily weighted at the lower end. This goes against the typical balance of ecology and biology in general which is usually slightly skewed to the right (at the higher end). Quantitative parasitology, therefore, utilizes advanced biostatistical methods and applications to understand such distributions.

Parasite Taxonomy and Phylogenetics.

Taxonomy: The science of classifying biological species.

Phylogeny: the examination of a species evolutionary history and lineage.

In simple terms, parasite taxonomy and phylogenetics are areas of study that concern the evolution of parasites in terms of their relationships to other species, common ancestry and their classification in the phylogenetic tree. This is a complex area prone to complications and confusion as few species (and this applies to botany and zoology broadly, not just parasitology) fit neatly in a single box. Relationships between species are complex because of shared genetic traits and independent traits evolving in separate species. Where parasitology is concerned, it is not just concerned with evolution and relationships between related species, but also in the evolution of species to their hosts - why they affect some and not others.

Veterinary Parasitology: Veterinary sciences concern the interactions between animals and humans that are an every-day part of human life. This includes pets and livestock, for example. Many of these species have the potential to be infected with a parasite and experience disease and disease symptoms when infected. Veterinary parasitology is the study of parasites in animals - mostly species used in a domestic context, but also in some areas of wildlife. As well as the interactions, researchers study the genesis and the development in animal hosts, of their parasites, and the taxonomy (see above), morphology, systematics, and life cycle. This is largely an area of medical science overlapping with treatment and vaccine research to ensure minimal impact of sickness on human health and economics and livelihood. Its primary function is to improve and protect health.

Current Parasitology Challenges and Developments

Climate Change Spreading Disease

It's been widely reported that a changing climate is leading to increased instances of some parasite-related diseases and some spreading out of their traditional tropical and subtropical zones. Diseases are already spreading thanks to the warming climate. Malaria cannot survive at temperatures below 20C and as climate change continues to increase warming some researchers expect malaria to spread to North Africa and eventually into southern Europe, similarly with dengue in the Pacific and Caribbean. Some expect it to reach the southern US states if temperatures continue to rise at the rate that they are presently rising. Evidence for this is limited at present, but temperature rises have yet to reach the required averages in the subtropical areas of the Americas, Africa, and southern Europe.

Climate Change Impact on Vulnerable Species

There are hundreds of species on the vulnerable and at-risk register. As parasites depend on these species, their continued existence - and the balance to the ecology that they provide - depend heavily on the existence of their hosts. Despite that many parasites such as those that cause dengue and malaria can expect to benefit from a changing climate, there are many more that are presently at risk and that is not good for their survival. Parasites are likely to become extinct before their host species. With limited numbers of hosts come limited chances for infection and therefore, parasite reproduction. One example of this is the marsupial woylie, also known as the brush-tailed bettong. Just one hundred years ago, it was endemic to Western Australia but with only 5,600 of their numbers, is on the brink of extinction. The situation for its most common parasite is even direr.

Prejudice Towards Parasitism

While researchers understand the importance of parasites to ecology and the food chain, that understanding has not transmitted well outside of the biological research arena. The very word "parasite" has a negative connotation; it is used against people and targeted groups. This continues the prejudice against these remarkable and unique plants and animals, fueling a belief that they not only don't serve a purpose - but also that they provide no benefit, taking and not giving. This has led to low funding and low interest in parasite research outside of the academic community for academia's sake. There is funding for medical research, but not simply to understand lifecycle and genetics. This has led to conservation issues of some critically endangered or vulnerable species.

Emerging Parasites

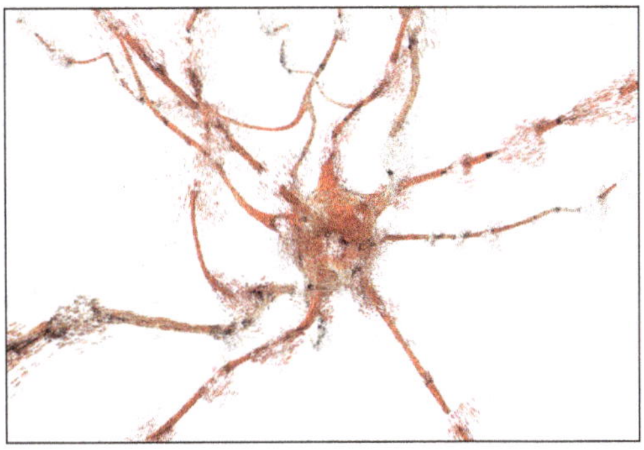

Because of parasite prevalence, diseases that they can cause is a global health concern. It is complex and resource-intensive to cope with the diseases that we know about. Climate change, resistance to treatment, remission or disappearance of other conditions and a number of other factors is leading to emerging or re-emerging diseases. trypanosomiasis, schistosomiasis, and leishmaniasis have all been in remission for generations but trypanosomiasis (resulting from the infection of a protozoan that spreads sleeping sickness to humans) is now in epidemic proportions in Africa in the last few years. Leishmaniasis, a New World disease spread by a protozoan called Leishmania spread by sand flies) is reaching out into new areas that had not previously experienced them.

Malaria is no emerging but it is spreading and now kills more people than all other diseases except tuberculosis.

VIROLOGY

Virology is the study of viruses and virus-like agents, including (but not limited to) their taxonomy, disease-producing properties, cultivation and genetics. It is often considered a part of microbiology or pathology. In the early years this discipline was dependent upon advances in the chemical and physical sciences, but viruses soon became tools for probing basic biochemical processes of cells.

Viruses have traditionally been viewed in a rather negative context as agents responsible for disease that must be controlled or eliminated. However, viruses also have certain beneficial properties that can be exploited for useful purposes (for example in gene therapy or vaccinology).

Characteristics and Classification of Viruses

Following the initial operational definition of a virus as a filterable agent, attempts were made to identify properties of viruses that separated them from other microorganisms. The single defining feature of all viruses is that they are obligate intracellular molecular parasites.

A second inviolate property of viruses is that they do not reproduce by binary fission – a method of asexual reproduction where pre-existing cells split into two identical daughter cells. For viruses, the process of reproduction is akin to an assembly line in which different parts come together to create new viral particles.

In general, viruses contain only one type of nucleic acid (either DNA or RNA) that carries the information necessary for viral replication. Nevertheless, it is clear now that some viruses contain other nucleic acid molecules; for example, in retroviruses, cellular transfer RNAs are essential for the action of the enzyme reverse transcriptase.

The chemical composition of viruses varies between different virus families. For the simplest of viruses, the virion is composed of viral structural proteins and nucleic acid, but the situation becomes more complex with the enveloped viruses. The latter types of viruses are maturing by budding through different cellular membranes that are modified by the insertion of viral proteins.

Several properties should be considered most important in constructing a scheme for classification of all the viruses: the nature of the nucleic acid present in the virion, the symmetry of the protein shell, dimensions of the virus particle and presence or absence of a lipid membrane.

The International Committee on Viral Taxonomy (ICTV), which was given the task of developing a universal taxonomic scheme for all the viruses, has put an emphasis on the viral genome (which is a blueprint for producing new viruses) as a basis of all classification decisions.

In formal virus taxonomy, families, subfamilies and genera are always written in italics, with the first letters of the names capitalized. Instead of formal names (e.g. Parvoviridae), common names are often used for viruses, even in medical literature (e.g. parvoviruses).

Study of Viruses

Even from the earliest times, it was clear that the filterable agents could not be cultivated on artificial media; this particular characteristic has withstood the test of time. Virus isolation in cell culture is still considered the gold standard against which other assays must be compared.

Still, the most obvious method of virus detection and identification is a direct visualization of the agent. The morphology of most viruses is sufficiently characteristic to identify the image as a virus and to assign an unknown virus to the adequate family. Furthermore, certain non-cultivable viruses can be detectable by electron microscopy.

The culture of animal cells typically involves the use of culture medium containing salts, glucose, vitamins, amino acids, antimicrobial drugs, buffers and (usually) blood serum which provides a source of necessary cellular growth factors. For certain cell-lines, defined serum-free media have been developed, which contain specific growth factors.

Serological tests are used to show the presence or absence of antibody to a specific virus. The presence of antibody indicates exposure to the agent, which may be due to a current clinical condition or to an earlier unrelated infection. Some examples are hemagglutination, complement fixation tests, radioimmunoassays, immunofluorescence, enzyme-linked immunosorbent assay (ELISA), radioimmune precipitation and Western blot assays.

Molecular techniques (such as polymerase chain reaction or PCR) rapidly gained acceptance and today have a myriad of applications. They are widely used for diagnostic clinical virology, as well as for research purposes. Such nucleic acid-centered technology also offered substantial advances in detection of viruses involving nucleic acid hybridization techniques.

MICROBIAL ECOLOGY

Microbial ecology (or environmental microbiology) is the ecology of microorganisms: their relationship with one another and with their environment. It concerns the three major domains of life—Eukaryota, Archaea, and Bacteria—as well as viruses.

Microorganisms, by their omnipresence, impact the entire biosphere. Microbial life plays a primary role in regulating biogeochemical systems in virtually all of our planet's environments, including some of the most extreme, from frozen environments and acidic lakes, to hydrothermal vents at the bottom of deepest oceans, and some of the most familiar, such as the human small intestine. As a consequence of the quantitative magnitude of microbial life (calculated as 5.0×1030 cells; eight orders of magnitude greater than the number of stars in the observable universe) microbes, by virtue of their biomass alone, constitute a significant carbon sink. Aside from carbon fixation, microorganisms' key collective metabolic processes (including nitrogen fixation, methane metabolism, and sulfur metabolism) control global biogeochemical cycling. The immensity of microorganisms' production is such that, even in the total absence of eukaryotic life, these processes would likely continue unchanged.

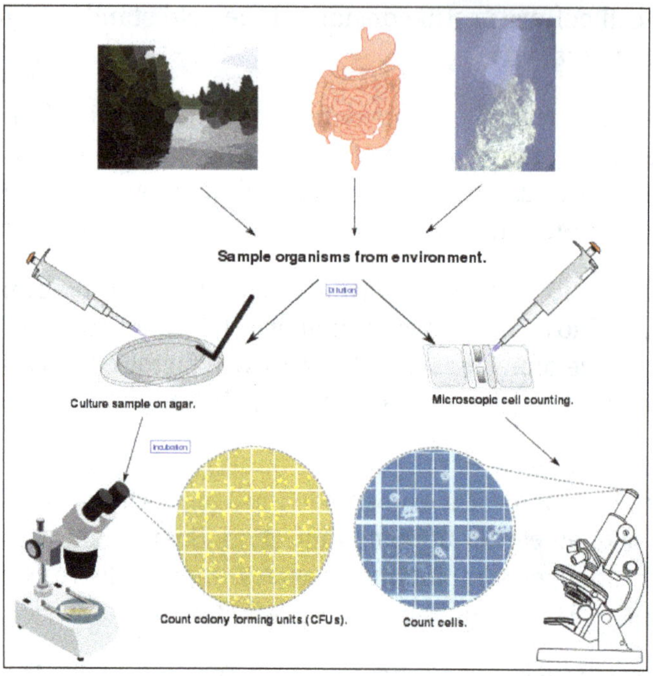

The great plate count anomaly. Counts of cells obtained via cultivation are orders of magnitude lower than those directly observed under the microscope. This is because microbiologists are able to cultivate only a minority of naturally occurring microbes using current laboratory techniques, depending on the environment.

While microbes have been studied since the seventeenth-century, this research was from a primarily physiological perspective rather than an ecological one. For instance, Louis Pasteur and his disciples were interested in the problem of microbial distribution both on land and in the ocean. Martinus Beijerinck invented the enrichment culture, a fundamental method of studying microbes from the environment. He is often incorrectly credited with framing the microbial biogeographic idea that "everything is everywhere, but, the environment selects", which was stated by Lourens Baas Becking. Sergei

Winogradsky was one of the first researchers to attempt to understand microorganisms outside of the medical context—making him among the first students of microbial ecology and environmental microbiology—discovering chemosynthesis, and developing the Winogradsky column in the process.

Beijerinck and Windogradsky, however, were focused on the physiology of microorganisms, not the microbial habitat or their ecological interactions. Modern microbial ecology was launched by Robert Hungate and coworkers, who investigated the rumen ecosystem. The study of the rumen required Hungate to develop techniques for culturing anaerobic microbes, and he also pioneered a quantitative approach to the study of microbes and their ecological activities that differentiated the relative contributions of species and catabolic pathways.

Roles

Microorganisms are the backbone of all ecosystems, but even more so in the zones where photosynthesis is unable to take place because of the absence of light. In such zones, chemosynthetic microbes provide energy and carbon to the other organisms.

Other microbes are decomposers, with the ability to recycle nutrients from other organisms' waste products. These microbes play a vital role in biogeochemical cycles. The nitrogen cycle, the phosphorus cycle, the sulphur cycle and the carbon cycle all depend on microorganisms in one way or another. For example, the nitrogen gas which makes up 78% of the earth's atmosphere is unavailable to most organisms, until it is converted to a biologically available form by the microbial process of nitrogen fixation.

Due to the high level of horizontal gene transfer among microbial communities, microbial ecology is also of importance to studies of evolution.

Symbiosis

Microbes, especially bacteria, often engage in symbiotic relationships (either positive or negative) with other microorganisms or larger organisms. Although physically small, symbiotic relationships amongst microbes are significant in eukaryotic processes and their evolution. The types of symbiotic relationship that microbes participate in include mutualism, commensalism, parasitism, and amensalism, and these relationships affect the ecosystem in many ways.

Mutualism

Mutualism in microbial ecology is a relationship between microbial species and between microbial species and humans that allow for both sides to benefit. One such example would be syntrophy, also known as cross-feeding, which is clearly shown in *Methanobacterium omelianskii*. Although initially thought of as one microbial species, this system is actually two species - an S organism and *Methabacterium bryantii*. The

S organism provides the bacterium with the H_2, which the bacterium needs in order to grow and produce methane. The reaction used by the S organism for the production of H_2 is endergonic (and so thermodynamically unfavored) however, when coupled to the reaction used by *Methabacterium bryantii* in its production of methane, the overall reaction becomes exergonic. Thus the two organisms are in a mutualistic relationship which allows them to grow and thrive in an environment, deadly for either species alone. Lichen is an example of a symbiotic organism.

Amensalism

Amensalism (also commonly known as antagonism) is a type of symbiotic relationship where one species/organism is harmed while the other remains unaffected. One example of such a relationship that takes place in microbial ecology is between the microbial species *Lactobacillus casei* and *Pseudomonas taetrolens*. When co-existing in an environment, *Pseudomonas taetrolens* shows inhibited growth and decreased production of lactobionic acid (its main product) most likely due to the byproducts created by *Lactobacillus casei* during its production of lactic acid. However, *Lactobacillus casei* shows no difference in its behaviour, and such this relationship can be defined as amensalism.

Microbial Resource Management

Biotechnology may be used alongside microbial ecology to address a number of environmental and economic challenges. For example, molecular techniques such as community fingerprinting can be used to track changes in microbial communities over time or assess their biodiversity. Managing the carbon cycle to sequester carbon dioxide and prevent excess methanogenesis is important in mitigating global warming, and the prospects of bioenergy are being expanded by the development of microbial fuel cells. Microbial resource management advocates a more progressive attitude towards disease, whereby biological control agents are favoured over attempts at eradication. Fluxes in microbial communities has to be better characterized for this field's potential to be realised. In addition, there are also clinical implications, as marine microbial symbioses are a valuable source of existing and novel antimicrobial agents, and thus offer another line of inquiry in the evolutionary arms race of antibiotic resistance, a pressing concern for researchers.

In Built Environment and Human Interaction

Microbes exist in all areas, including homes, offices, commercial centers, and hospitals. In 2016, the journal *Microbiome* published a collection of various works studying the microbial ecology of the built environment.

A 2006 study of pathogenic bacteria in hospitals found that their ability to survive varied by the type, with some surviving for only a few days while others survived for months.

The lifespan of microbes in the home varies similarly. Generally bacteria and viruses require a wet environment with a humidity of over 10 percent. *E. coli* can survive for a few hours to a day. Bacteria which form spores can survive longer, with *Staphylococcus aureus* surviving potentially for weeks or, in the case of *Bacillus anthracis*, years.

In the home, pets can be carriers of bacteria; for example, reptiles are commonly carriers of salmonella.

S. aureus is particularly common, and asymptomatically colonizes about 30% of the human population; attempts to decolonize carriers have met with limited success and generally involve mupirocin nasally and chlorhexidine washing, potentially along with vancomycin and cotrimoxazole to address intestinal and urinary tract infections.

Antimicrobials

Some metals, particularly copper and silver, have antimicrobial properties. Using antimicrobial copper-alloy touch surfaces is a technique which has begun to be used in the 21st century to prevent transmission of bacteria. Silver nanoparticles have also begun to be incorporated into building surfaces and fabrics, although concerns have been raised about the potential side-effects of the tiny particles on human health.

MICROBIAL GENETICS

Microbial genetics is a branch of genetics concerned with the transmission of hereditary characters in microorganisms . Within the usual definition, microorganisms include prokaryotes like bacteria , unicellular or mycelial eukaryotes e.g., yeasts and other fungi , and viruses , notably bacterial viruses (bacteriophages). Microbial genetics has played a unique role in developing the fields of molecular and cell biology and also has found applications in medicine, agriculture, and the food and pharmaceutical industries.

Because of their relative simplicity, microbes are ideally suited for combined biochemical and genetic studies, and have been successful in providing information on the genetic code and the regulation of gene activity. The operon model formulated by French biologists François Jacob and Jacques Monod in 1961, is one well known example. Based on studies on the induction of enzymes of lactose catabolism in the bacterium Escherichia coli, the operon has provided the groundwork for studies on gene expression and regulation, even up to the present day. The many applications of microbial genetics in medicine and the pharmaceutical industry emerge from the fact that microbes are both the causes of disease and the producers of antibiotics . Genetic studies have been used to understand variation in pathogenic microbes and also to increase the yield of antibiotics from other microbes.

Hereditary processes in microorganisms are analogous to those in multicellular organisms. In both prokaryotic and eukaryotic microbes, the genetic material is DNA ; the only known exceptions to this rule are the RNA viruses. Mutations , heritable changes in the DNA, occur spontaneously and the rate of mutation can be increased by mutagenic agents. In practice, the susceptibility of bacteria to mutagenic agents has been used to identify potentially hazardous chemicals in the environment. For example, the Ames test was developed to evaluate the mutagenicity of a chemical in the following way. Plates containing a medium lacking in, for example, the nutrient histidine are inolculated with a histidine requiring strain of the bacterium Salmonella typhimurium. Thus, only cells that revert back to the wild type can grow on the medium. If plates are exposed to a mutagenic agent, the increase in the number of mutants compared with unexposed plates can be observed and a large number of revertants would indicate a strong mutagenic agent. For such studies, microorganisms offer the advantage that they have short mean generation times, are easily cultured in a small space under controlled conditions and have a relatively uncomplicated structure.

Microorganisms, and particularly bacteria, were generally ignored by the early geneticists because of their small in size and apparent lack of easily identifiable variable traits. Therefore, a method of identifying variation and mutation in microbes was fundamental for progress in microbial genetics. As many of the mutations manifest themselves as metabolic abnormalities, methods were developed by which microbial mutants could be detected by selecting or testing for altered phenotypes. Positive selection is defined as the detection of mutant cells and the rejection of unmutated cells. An example of this is the selection of penicillin resistant mutants, achieved by growing organisms in media containing penicillin such that only resistant colonies grow. In contrast, negative selection detects cells that cannot perform a certain function and is used to select mutants that require one or more extra growth factors. Replica plating is used for negative selection and involves two identical prints of colony distributions being made on plates with and without the required nutrients. Those microbes that do not grow on the plate lacking the nutrient can then be selected from the identical plate, which does contain the nutrient.

The first attempts to use microbes for genetic studies were made in the USA shortly before World War II, when George W. Beadle and Edward L. Tatum employed the fungus, Neurospora, to investigate the genetics of tryptophan metabolism and nicotinic acid synthesis. This work led to the development of the "one gene one enzyme" hypothesis. Work with bacterial genetics, however, was not really begun until the late 1940s. For a long time, bacteria were thought to lack sexual reproduction, which was believed to be necessary for mixing genes from different individual organisms—a process fundamental for useful genetic studies. However, in 1947, Joshua Lederberg working with Edward Tatum demonstrated the exchange of genetic factors in the bacterium, Escherichia coli. This process of DNA transfer was termed conjugation and requires cell-to-cell contact between two bacteria. It is controlled by genes carried by plasmids , such as the fertility (F) factor, and typically involves the transfer of the plasmid from donor torecipient cell.

Other genetic elements, however, including the donor cell chromosome, can sometimes also be mobilized and transferred. Transfer to the host chromosome is rarely complete, but can be used to map the order of genes on a bacterial genome.

Other means by which foreign genes can enter a bacterial cell include transformation , transfection, and transduction . Of the three processes, transformation is probably the most significant. Evidence of transformation in bacteria was first obtained by the British scientist, Fred Griffith in the late 1920s working with Streptococcus pneumoniae and the process was later explained in the 1930s by Oswald Avery and his associates at the Rockefeller Institute in New York. It was discovered that certain bacteria exhibit competence, a state in which cells are able to take up free DNA released by other bacteria. This is the process known as transformation, however, relatively few microorganisms can be naturally transformed. Certain laboratory procedures were later developed that make it possible to introduce DNA into bacteria, for example electroporation, which modifies the bacterial membrane by treatment with an electric field to facilitate DNA uptake. The latter two processes, transfection and transduction, involve the participation of viruses for nucleic acid transfer. Transfection occurs when bacteria are transformed with DNA extracted from a bacterial virus rather than from another bacterium. Transduction involves the transfer of host genes from one bacterium to another by means of viruses. In generalized transduction, defective virus particles randomly incorporate fragments of the cell DNA; virtually any gene of the donor can be transferred, although the efficiency is low. In specialized transduction, the DNA of a temperate virus excises incorrectly and brings adjacent host genes along with it. Only genes close to the integration point of the virus are transduced, and the efficiency may be high.

After the discovery of DNA transfer in bacteria, bacteria became objects of great interest to geneticists because their rate of reproduction and mutation is higher than in larger organisms; i.e., a mutation occurs in a gene about one time in 10,000,000 gene duplications, and one bacterium may produce 10,000,000,000 offspring in 48 hours. Conjugation, transformation, and transduction have been important methods for mapping the genes on the chromosomes of bacteria. These techniques, coupled with restriction enzyme analysis, cloning DNA sequencing, have allowed for the detailed studies of the bacterial chromosome. Although there are few rules governing gene location, the genes encoding enzymes for many biochemical pathways are often found tightly linked in operons in prokaryotes. Large scale sequencing projects revealed the complete DNA sequence of the genomes of several prokaryotes, even before eukaryotic genomes were considered.

CELLULAR MICROBIOLOGY

Cellular microbiology is a discipline that bridges microbiology and cell biology.

The term "cellular microbiology" was coined by the authors of the book of the same title published in 1996. Cooperation and mutual dependency between microbiology and cell biology had been increasing in the years before that, and the emergence of a new discipline had been suggested and discussed in several scientific conferences.

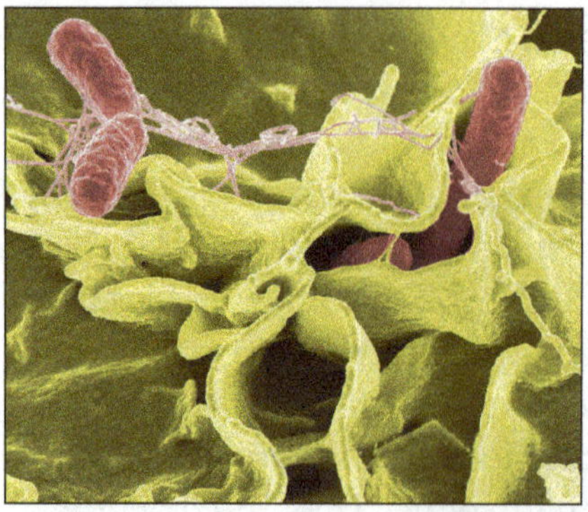

Salmonella bacteria (red) invade cultured human cells.

Cellular microbiology attempts to use pathogenic microorganisms as tools for cell-biology research, and to employ cell-biology methods to understand the pathogenicity of microorganisms. Toxins and virulence factors from microbes have been used for decades to influence processes in eukaryotic cells and to study them. It has increasingly appeared that applying a purified toxin on a cell does not always provide the complete picture, and that understanding the role of the toxin in pathogenicity, the way the toxin promotes the microbe, the way the toxin is produced and the co-evolution of the toxin and its host-cell counterparts, is crucial.

Numerous eukaryotic cellular processes have been clarified using microbial "tools". A major subject in this category is the cytoskeleton. Many microbes modify and influence the synthesis or degradation of the host-cell cytoskeleton, in particular the actin network. Intracellular microbes, such as the bacteria *Salmonella* and *Shigella*, elicit actin polymerization in host cells that otherwise do not internalize microbes (non-phagocytes). This causes the formation of projections that eventually engulf the bacteria. Bacteria such as *Yersinia* inhibit actin polymerization in phagocytes, thereby preventing their uptake. Cellular microbiology tries to understand these processes and how they promote infection. Other eukaryotic processes that microbes influence and that are researched using microbes are signal transduction, metabolism, vesicle trafficking, cell cycle and transcriptional regulation, to name but a few.

Recently, the field of Cellular Microbiology has been expanded to incorporate investigation of the cell biology of microbes themselves. "The field of cellular microbiology is a coalescence of two fields: molecular microbiology and cell biology," said Professor Jacek Hawiger, Chair of Microbiology and Immunology at Vanderbilt University. Particularly in the case of bacterial cells, new technology is starting to be used to reveal a high level of organization within the bacterial cells themselves. For example, high-resolution fluorescence microscopy and atomic force microscopy are both being used to show just how sophisticated bacterial cells are.

MEDICAL MICROBIOLOGY

Medical microbiology , the large subset of microbiology that is applied to medicine, is a branch of medical science concerned with the prevention, diagnosis and treatment of infectious diseases. In addition, this field of science studies various clinical applications of microbes for the improvement of health. There are four kinds of microorganisms that cause infectious disease: bacteria, fungi, parasites and viruses, and one type of infectious protein called prion.

A microbiologist examining cultures under a dissecting microsope.

A medical microbiologist studies the characteristics of pathogens, their modes of transmission, mechanisms of infection and growth. Using this information, a treatment can be devised. Medical microbiologists often serve as consultants for physicians, providing identification of pathogens and suggesting treatment options. Other tasks may include the identification of potential health risks to the community or monitoring the evolution of potentially virulent or resistant strains of microbes, educating the community and assisting in the design of health practices. They may also assist in preventing or controlling epidemics and outbreaks of disease. Not all medical microbiologists study microbial pathology; some study common, non-pathogenic species to determine whether their properties can be used to develop antibiotics or other treatment methods.

Epidemiology, the study of the patterns, causes, and effects of health and disease conditions in populations, is an important part of medical microbiology, although the clinical aspect of the field primarily focuses on the presence and growth of microbial infections in individuals, their effects on the human body, and the methods of treating those infections. In this respect the entire field, as an applied science, can be conceptually subdivided into academic and clinical subspecialties, although in reality there is a fluid continuum between public health microbiology and clinical microbiology, just as the state of the art in clinical laboratories depends on continual improvements in academic medicine and research laboratories.

Commonly Treated Infectious Diseases

Bacterial:

- Streptococcal pharyngitis.
- Chlamydia.
- Typhoid fever.
- Tuberculosis.

Viral:

- Rotavirus.
- Hepatitis C.
- Human papillomavirus (HPV).

Parasitic:

- Malaria.
- *Giardia lamblia.*
- *Toxoplasma gondii.*

Fungal:

- Candida.
- Histoplasmosis.

Causes and Transmission of Infectious Diseases

Infections may be caused by bacteria, viruses, fungi, and parasites. The pathogen that causes the disease may be exogenous (acquired from an external source; environmental, animal or other people, e.g. Influenza) or endogenous (from normal flora e.g. candidiasis).

The site at which a microbe enters the body is referred to as the portal of entry. These include the respiratory tract, gastrointestinal tract, genitourinary tract, skin, and mucous membranes. The portal of entry for a specific microbe is normally dependent on how it travels from its natural habitat to the host.

There are various ways in which disease can be transmitted between individuals. These include:

- Direct contact: Touching an infected host, including sexual contact.
- Indirect contact: Touching a contaminated surface.

- Droplet contact: Coughing or sneezing.
- Fecal–oral route: Ingesting contaminated food or water sources.
- Airborne transmission: Pathogen carrying spores.
- Vector transmission: An organism that does not cause disease itself but transmits infection by conveying pathogens from one host to another.
- Fomite transmission: An inanimate object or substance capable of carrying infectious germs or parasites.
- Environmental: Hospital-acquired infection (Nosocomial infections).

Like other pathogens, viruses use these methods of transmission to enter the body, but viruses differ in that they must also enter into the host's actual cells. Once the virus has gained access to the host's cells, the virus' genetic material (RNA or DNA) must be introduced to the cell. Replication between viruses is greatly varied and depends on the type of genes involved in them. Most DNA viruses assemble in the nucleus while most RNA viruses develop solely in cytoplasm.

The mechanisms for infection, proliferation, and persistence of a virus in cells of the host are crucial for its survival. For example, some diseases such as measles employ a strategy whereby it must spread to a series of hosts. In these forms of viral infection, the illness is often treated by the body's own immune response, and therefore the virus is required to disperse to new hosts before it is destroyed by immunological resistance or host death. In contrast, some infectious agents such as the Feline leukemia virus, are able to withstand immune responses and are capable of achieving long-term residence within an individual host, whilst also retaining the ability to spread into successive hosts.

Diagnostic Tests

Identification of an infectious agent for a minor illness can be as simple as clinical presentation; such as gastrointestinal disease and skin infections. In order to make an educated estimate as to which microbe could be causing the disease, epidemiological factors need to be considered; such as the patient's likelihood of exposure to the suspected organism and the presence and prevalence of a microbial strain in a community.

Diagnosis of infectious disease is nearly always initiated by consulting the patient's medical history and conducting a physical examination. More detailed identification techniques involve microbial culture, microscopy, biochemical tests and genotyping. Other less common techniques (such as X-rays, CAT scans, PET scans or NMR) are used to produce images of internal abnormalities resulting from the growth of an infectious agent.

Microbial Culture

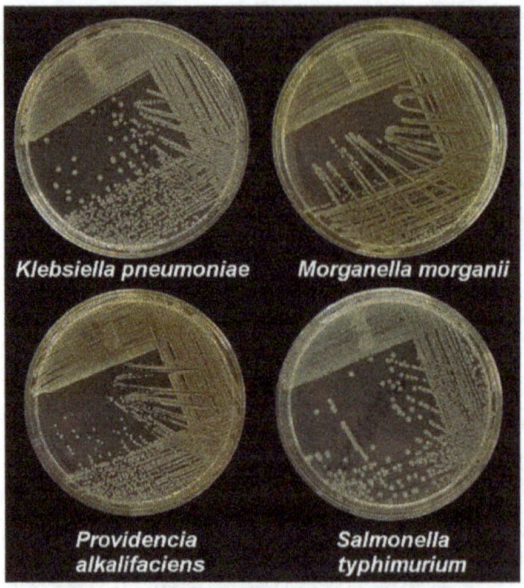

Four nutrient agar plates growing colonies of common Gram negative bacteria.

Microbiological culture is the primary method used for isolating infectious disease for study in the laboratory. Tissue or fluid samples are tested for the presence of a specific pathogen, which is determined by growth in a selective or differential medium.

The 3 main types of media used for testing are:

- Solid culture: A solid surface is created using a mixture of nutrients, salts and agar. A single microbe on an agar plate can then grow into colonies (clones where cells are identical to each other) containing thousands of cells. These are primarily used to culture bacteria and fungi.
- Liquid culture: Cells are grown inside a liquid media. Microbial growth is determined by the time taken for the liquid to form a colloidal suspension. This technique is used for diagnosing parasites and detecting mycobacteria.
- Cell culture: Human or animal cell cultures are infected with the microbe of interest. These cultures are then observed to determine the effect the microbe has on the cells. This technique is used for identifying viruses.

Microscopy

Culture techniques will often use a microscopic examination to help in the identification of the microbe. Instruments such as compound light microscopes can be used to assess critical aspects of the organism. This can be performed immediately after the sample is taken from the patient and is used in conjunction with biochemical staining techniques, allowing for resolution of cellular features. Electron microscopes and fluorescence microscopes are also used for observing microbes in greater detail for research.

Biochemical Tests

Fast and relatively simple biochemical tests can be used to identify infectious agents. For bacterial identification, the use of metabolic or enzymatic characteristics are common due to their ability to ferment carbohydrates in patterns characteristic of their genus and species. Acids, alcohols and gases are usually detected in these tests when bacteria are grown in selective liquid or solid media, as mentioned above. In order to perform these tests en masse, automated machines are used. These machines perform multiple biochemical tests simultaneously, using cards with several wells containing different dehydrated chemicals. The microbe of interest will react with each chemical in a specific way, aiding in its identification.

Serological methods are highly sensitive, specific and often extremely rapid laboratory tests used to identify different types of microorganisms. The tests are based upon the ability of an antibody to bind specifically to an antigen. The antigen (usually a protein or carbohydrate made by an infectious agent) is bound by the antibody, allowing this type of test to be used for organisms other than bacteria. This binding then sets off a chain of events that can be easily and definitively observed, depending on the test. More complex serological techniques are known as immunoassays. Using a similar basis as described above, immunoassays can detect or measure antigens from either infectious agents or the proteins generated by an infected host in response to the infection.

Polymerase Chain Reaction

Polymerase chain reaction (PCR) assays are the most commonly used molecular technique to detect and study microbes. As compared to other methods, sequencing and analysis is definitive, reliable, accurate, and fast. Today, quantitative PCR is the primary technique used, as this method provides faster data compared to a standard PCR assay. For instance, traditional PCR techniques require the use of gel electrophoresis to visualize amplified DNA molecules after the reaction has finished. quantitative PCR does not require this, as the detection system uses fluorescence and probes to detect the DNA molecules as they are being amplified. In addition to this, quantitative PCR also removes the risk of contamination that can occur during standard PCR procedures (carrying over PCR product into subsequent PCRs). Another advantage of using PCR to detect and study microbes is that the DNA sequences of newly discovered infectious microbes or strains can be compared to those already listed in databases, which in turn helps to increase understanding of which organism is causing the infectious disease and thus what possible methods of treatment could be used. This technique is the current standard for detecting viral infections such as AIDS and hepatitis.

Treatments

Once an infection has been diagnosed and identified, suitable treatment options must be assessed by the physician and consulting medical microbiologists. Some infections can be dealt with by the body's own immune system, but more serious infections are treated with antimicrobial drugs. Bacterial infections are treated with antibacterials (often called antibiotics) whereas fungal and viral infections are treated with antifungals and antivirals respectively. A broad class of drugs known as antiparasitics are used to treat parasitic diseases.

Medical microbiologists often make treatment recommendations to the patient's physician based on the strain of microbe and its antibiotic resistances, the site of infection, the potential toxicity of antimicrobial drugs and any drug allergies the patient has.

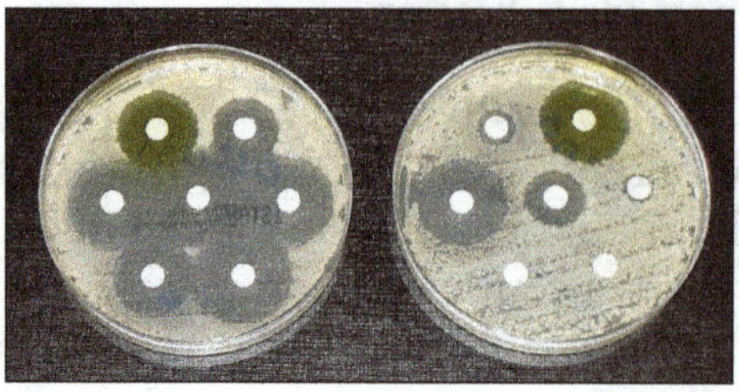

Antibiotic resistance tests: bacteria in the culture on the left are sensitive to the antibiotics contained in the white, paper discs. Bacteria in the culture on the right are resistant to most of the antibiotics.

In addition to drugs being specific to a certain kind of organism (bacteria, fungi, etc.), some drugs are specific to a certain genus or species of organism, and will not work on other organisms. Because of this specificity, medical microbiologists must consider the effectiveness of certain antimicrobial drugs when making recommendations. Additionally, strains of an organism may be resistant to a certain drug or class of drug, even when it is typically effective against the species. These strains, termed resistant strains, present a serious public health concern of growing importance to the medical industry as the spread of antibiotic resistance worsens. Antimicrobial resistance is an increasingly problematic issue that leads to millions of deaths every year.

Whilst drug resistance typically involves microbes chemically inactivating an antimicrobial drug or a cell mechanically stopping the uptake of a drug, another form of drug resistance can arise from the formation of biofilms. Some bacteria are able to form biofilms by adhering to surfaces on implanted devices such as catheters and prostheses and creating an extracellular matrix for other cells to adhere to. This provides them with a stable environment from which the bacteria can disperse and infect other parts of the host. Additionally, the extracellular matrix and dense outer layer of bacterial cells can protect the inner bacteria cells from antimicrobial drugs.

Medical microbiology is not only about diagnosing and treating disease, it also involves the study of beneficial microbes. Microbes have been shown to be helpful in combating infectious disease and promoting health. Treatments can be developed from microbes, as demonstrated by Alexander Fleming's discovery of penicillin as well as the development of new antibiotics from the bacterial genus Streptomyces among many others.

Not only are microorganisms a source of antibiotics but some may also act as probiotics to provide health benefits to the host, such as providing better gastrointestinal health or inhibiting pathogens.

DIAGNOSTIC MICROBIOLOGY

Diagnostic microbiology is the study of microbial identification. Since the discovery of the germ theory of disease, scientists have been finding ways to harvest specific organisms. Using methods such as differential media or genome sequencing, physicians and scientists can observe novel functions in organisms for more effective and accurate diagnosis of organisms. Methods used in diagnostic microbiology are often used to take advantage of a particular difference in organisms attain information about what species it might be, often through a reference of previous studies. New studies provide information that other scientists can reference back to so scientists can have a baseline knowledge of the organism he or she are working with.

Atmosphere and Length of Incubation

When culturing microbes, different organisms require different environments for optimal growing conditions.

Gas-Pak jar.

Aerobic vs. Anaerobic

Anaerobic organisms require an oxygen free environment. When culturing anaerobic microbes, broths are often flushed with nitrogen gas to extinguish oxygen present, and growth can also occur on media in a chamber without oxygen present. Sodium resazurin can be added to indicate redox potential. Cultures are to be incubated in an oxygen free environment for 48 hours at 35°C before growth is examined.

Anaerobic bacteria collection can come from a variety of sources in patient samples, including blood, bile, bone marrow, cerebrospinal fluid, direct lung aspirate, tissue biopsies from a normally sterile site, fluid from a normally sterile site (like a joint), dental, abscess, abdominal or pelvic abscess, knife, gunshot, or surgical wound, or severe burn.

Incubation Length

Incubation times vary based upon the microbe that requires culturing. Traditional culturing techniques, for example, require less than 24 hours culture time for *Escherichia coli* but 6–8 weeks for successful culturing of *Mycobacterium tuberculosis* before definitive results are expressed. A benefit of non-culture tests is that physicians and microbiologists are not handicapped by waiting periods.

Incubation follows a growth curve variable for every microorganism. Cultures follow a lag, log, stationary, and finally death phase. The lag phase is not well known in microbiology, but it is speculated that this phase consists of the microorganism adjusting to its environment by synthesizing proteins specific for the surrounding habitat. The log phase is the period where a culture experiences logarithmic growth until nutrients become scarce. The stationary phase is when culture concentration is the highest and cells stop reproducing. When nutrients in the environment are depleting, organisms enter the death phase where toxic metabolites become abundant and nutrients are depleted to the point where cell death exceeds reproduction.

Rapid Identification after Culture

Automated Culturing Systems

Automatic cell culturing systems are becoming popular because of their ability to maintain a sterile growth environment and remove strain on the laboratory staff involving repetitive experimentation. Laboratories can also set incubation times to adjust for the lag period involved in bacterial growth.

Blood Cultures

Blood cultures can allow for diagnostic results after culture. Recent development of DNA based PCR diagnostics have provided faster diagnostic results as opposed to overnight biochemical tests. DNA diagnostic test can diagnose with near the same specificity as biochemical test, resulting in the same diagnostic result in 90% of cases.

Breath Tests

Breath test for microbial diagnosis on patients has been used in a clinical setting for bacteria, including *Helicobacter pylori*. Diagnostic test using the breath of patients look for metabolites excreted that were manufactured by the infectious microorganism.

H. pylori is tested by testing patients for CO_2 concentration, increased because of the organism's ability to convert urea into other derivatives.

Conventional Tests

Antibody Detection

A benefit of antibody detection (ELISA) is that protein identification on a microorganism becomes faster than a western blot. Antibody detection works by attaching an indicator to an antibody with a known specificity and observing whether the antibody attaches. ELISA can also indicate viral presence and is highly specific, having a detection specificity of 10^{-9}-10^{-12} moles per litre detection. By knowing the epitope sequence of the antibody, ELISA can also be used for antigen detection in a sample.

Histological Detection and Culture

Histological methods used for microbiology are useful because of their ability to quickly identify a disease present in a tissue biopsy.

Rapid Antigen Tests

Immunofluorescence

Immunofluorescence is performed by the production of anti-antibodies with a fluorescent molecule attached, making it a chemiluminescent molecule, which provides a glow when subject to ultraviolet light. Antibodies are added to a bacterial solution, providing an antigen for the binding of fluorescent anti-antibody adherence.

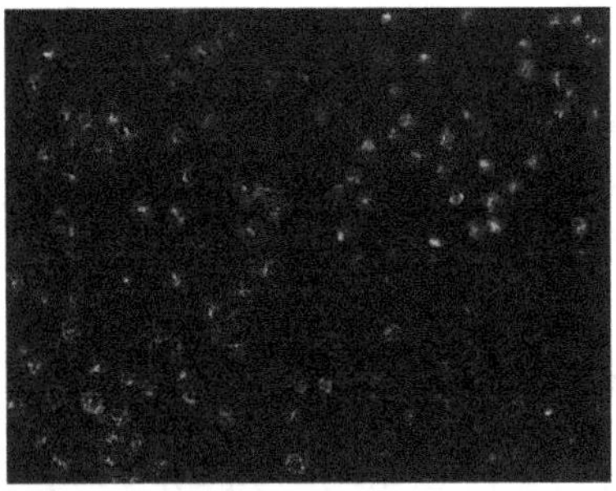

Immunofluorescence.

Mass Spectrometry

Mass spectrometry is a clinical application involving diagnosis of bacteria-specific molecules. By using a known database of previously catalogued organisms, similar techniques can produce results that reference back to this database, resulting in rapid diagnosis of microorganisms. Recent studies have suggested that these tests can become specific enough to diagnose down to the sub-species level by observing novel biomarkers.

Biochemical Profile-based Microbial Identification Systems

Biochemical tests are used to identify microbes based on biochemical pathways present in those microbes.

6.5% Salt Broth

The 6.5% salt broth test is used to analyze the tolerance level of various bacteria under halophilic conditions. This test is used because most organisms cannot survive in high salt concentrations while *Staphylococci*, *Enterococci*, and *Aerococci* are all expected to tolerate 6.5% NaCl concentrations.

Acetate Utilization

The acetate utilization test is used primarily to differentiate between *Escherichia coli* from members of the genus *Shigella*. Many of the *Escherichia coli* strains have the capability of utilization of acetate for a sole carbon and energy source, while *Shigella* does not. Since acetate utilization results in an increase in pH, an indicator is added that changes color under conditions of acetate utilization.

ALA

An ALA (delta-aminolevulinic acid) test is used to test for the presence of porphyrin and cytochrome compounds. Finding hemin synthesis indicates that the organism is likely *Haemophilus*.

Aminopeptidase

The aminopeptidase test analyzes bacteria for the production of the enzyme L-alanine-aminopeptidase, an enzyme found in many gram-negative bacteria. Adding L-Alanine-4-nitroanilide hydrochloride to a bacterial culture works as an indicator, changing to a yellow color in the presence of L-alanine-aminopeptidase.

Analytical Profile Index

An analytical profile index is a fast identification system based on biochemical incubation tests. Usually, this test is used to quickly diagnose clinically relevant bacteria by allowing physicians to run about 20 tests at one time.

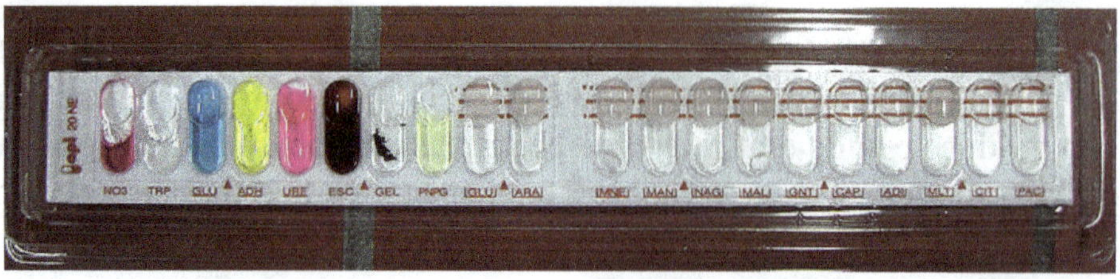

Antibiotic Disks

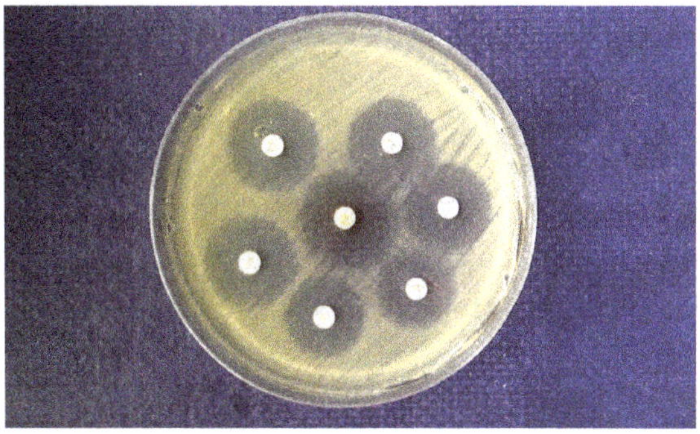

Antibiotic susceptible bacteria.

Antibiotic disks are used to test the ability for an antibiotic to inhibit growth of a microorganism. This method, which is commonly used with Mueller-Hinton agar, is used by evenly seeding bacteria over a petri dish and applying an antibiotic treated disk to the top of the agar. By observing the ring formed around the disk formed due to the lack of bacterial growth, the zone of inhibition can be found, which is used to find the susceptibility of an organism to an antibiotic.

Bile Esculin Agar

The bile esculin test is used to differentiate members of the genus *Enterococcus* from *Streptococcus*.

Bile Solubility

Bile solubility is used to test for *Staphylococcus Pneumoniae* due to their unique ability to be lysed by sodium deoxycholate. Lysis indicates *S. Pneumoniae* while no lysis does not.

Camp

A CAMP test is used to differentiate between *Streptococcus agalactiae* and other species of beta-hemolytic *Streptococcus*. This biochemical test uses the fact that *Streptococcus agalactiae* excretes a CAMP substance, making it slightly more hemolytic, which can be observed on blood agar media.

Catalase

The catalase test tests whether a microbe produce the enzyme catalase, which catalyzes the breakdown of hydrogen peroxide. Smearing a colony sample onto a glass slide and adding a solution of hydrogen peroxide will indicate whether the enzyme is present or not. Bubbling is a positive test while nothing happening is a negative result.

Cetrimide Agar

Cetrimide agar slants is a selective agar used to isolate *Pseudomonas aeruginosa.*

CLO Tests

The CLO test is used to diagnose *H. Pylori* in patient biopsies. A sample of the biopsy is places in a medium containing urea, which *H. Pylori* can use in some of its biochemical pathways. Consumption of urea indicates a positive test result.

Coagulase

The coagulase test determines whether an organism can produce the enzyme coagulase, which causes the fibrin to clot. Inoculating a plasma test tube with the microbe indicates whether coagulase is produced. A clot indicates the presence of coagulase, while no clot indicates the lack of coagulase.

DNA Hydrolysis

DNase agar is used to test whether a microbe can produce the exoenzyme deoxyribonuclease (DNase), which hydrolyzes DNA. Methyl green is used as an indicator in the growth medium because it is a cation that is provides an opaqueness to a medium with the presence of negatively charged DNA strands. When DNA is cleaved, the media becomes clear, showing the presence of DNase activity.

Gelatin

The gelatin test is used to analyze whether a microbe can hydrolyze gelatin with the enzyme gelatinase. The gelatin makes the agar solid, so if an organism can produce gelatinase and consume gelatin as an energy and carbon source, the agar will become liquid during growth.

Gonocheck II

The Gonochek II test, a commercial biochemical test, is used to differentiate between *Neisseria lactamica, Neisseria meningitidis, N. gonorrhoeae* and *Moraxella catarrhalis.* The principle behind this test is to use enzymes native to the organism to create a colored product in the presence of foreign molecules. The chemical [[5-bromo-4-chloro-3-indolyl-beta-D-galactoside is used in the test because *N. lactamica* can hydrolyze it with the production of β-galactosidase, turning the solution into a blue color. Gamma-glutamyl-p-nitroanilide is added to the solution to indicate whether the bacteria is *N. meningitides,* which hydrolyzes the molecule with the enzyme gamma-glutamylaminopeptidase, producing a yellow end-product. Prolyl-4-methoxynaphthylamide is in the solution to identify *N. gonorrhoeae* because of its ability to hydrolyze the molecule with the enzyme hydroxyprolylaminopeptidase, creating a red-pink derivative. *M. catarrhalis* contains none of these enzymes, rendering the solution colorless. This process of identification takes approximately 30 minutes in total.

Hippurate

The Hippurate diagnostic test is used to differentiate between *Gardnerella vaginalis, Campylobacter jejuni, Listeria monocytogenes* and group B streptococci using the chemical Hippurate. The Hippurate hydrolysis pathway, capable by organisms with the necessary enzymes, produces glycine as a byproduct. Using the indicator ninhydrin, which changes color in the presence of glycine, will display either a colorless product, a negative result, of a dark blue color, a positive result.

Indole Butyrate Disk

An indole butyrate disc is used to differentiate between *Neisseria gonorrhoeae* (negative result) and *Moraxella catarrhalis* (positive result). This test involves a butyrate disk, which when smeared with a culture, will change color for a positive result after 5 minutes of incubation. A blue color is the result of a positive test.

Lysine Iron Agar Slant

The lysine iron agar slant test is used to tell whether an organism can decarboxylate lysine and/or produce hydrogen sulfide.

Lysostaphin

The lysostaphin test is used to differentiate between *Staphylococcus* and *Micrococcus* bacteria. Lysostaphin can lyse *Staphylococcus,* but *Micrococcus* bacteria are resistant to the chemical.

Methyl Red Test

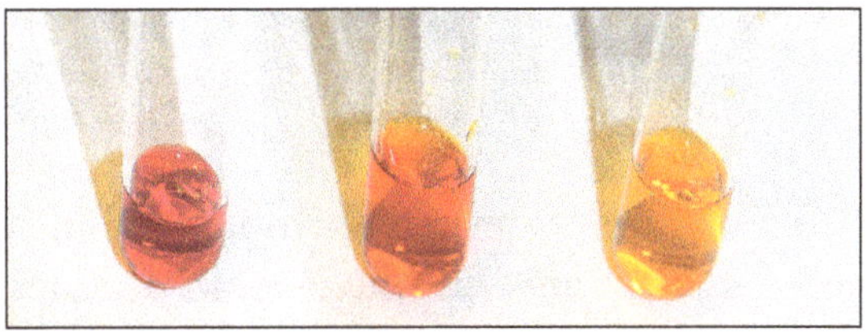

Color transition of Methyl red solution under different acid-base conditions.

The methyl red test is used to analyze whether a bacterium produces acids through sugar fermentation.

Microdase

Microdase is a modified oxidase test used to differentiate *Micrococcus* from

Staphylococcus by testing for the presence of cytochrome c. A positive result produces a dark color around the inoculant while negative result produces no color change.

Nitrite Test

The nitrite test is commonly used to diagnose urinary tract infections by measuring the concentrations of nitrite in solution, indicating the presence of a gram-negative organism.

Oxidase

The oxidase test indicates whether a microbe is aerobic. By using the chemical N,N,N,N-tetramethyl-1,4-phenylendiamin, an electron acceptor that changes color when oxidized by cytochrome c oxidase, one can deduce whether a microbe can perform aerobic respiration. A color change to purple indicates oxidative respiration while no color change provides evidence that the organism does not have cytochrome c oxidase.

Phenylalanine Deaminase

The phenylalanine deaminase test is used to tell whether an organism can produce the enzyme deaminase. Deaminase is the enzyme that can deaminate the amino acid phenylalanine into the products ammonia and phenylpyruvic acid. The test is performed by adding phenylalanine to the growth medium and allowing growth to occur. After incubation, 10% ferric chloride is added to the solution, which will react with phenylpyruvic acid in solution to make a dark green color, resulting in a positive test result.

PYR

The PYR test is used to check if an organism has enzymes to hydrolyze L-pyrrolidonyl-β-napthylamide. A positive result indicates that the organism is either group A *streptococcus* and/or group D *enterococcus*.

Reverse CAMP

The reverse CAMP test utilizes the synergetic hemolytic abilities of the CAMP factor produced by *Streptococcus agalactiae* with the α-toxin produced by *Clostridium perfringens*. Streaking these two organisms perpendicular to each other on a blood agar plate will yield a "bow-tie" clearing of the blood agar by the hemolytic capabilities of the two organisms' toxins. Incubation requires 24 hours at 37°C.

Simmons' Citrate Agar

Simmons' citrate agar is used to test whether an organism can utilize citrate for its sole carbon source.

Spot Indole

The spot indole test is used to determine if a microbe can deaminate tryptophan to produce indole. This test is performed by saturating a piece of filter paper with Indole Kovacs Reagent and scraping a portion of microbe onto the paper. A color to a pink-red color indicates a positive result while no color change indicates the lack of tryptophanase.

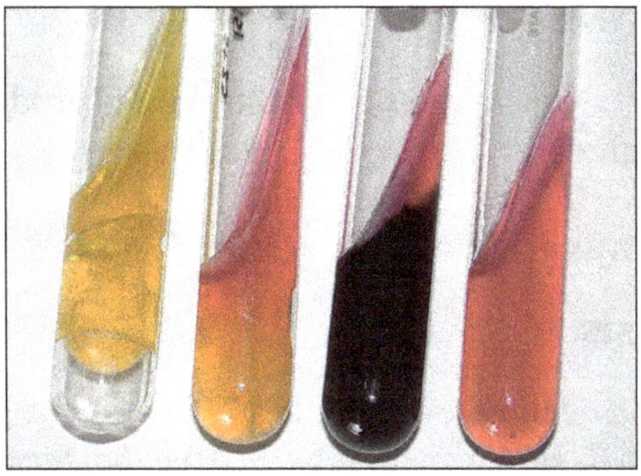

TSI Agar.

Sulphide Indole Motility Medium

The sulfide indole motility medium is a three-part test for an organism's ability to reduce sulfates, produce indoles, and motile ability.

TSI Slant

The triple sugar iron (TSI) test is a differential media used to tell whether an organism can ferment glucose, sucrose, and/or lactose and whether an organism can produce hydrogen sulfide gas.

Urea Agar Slant

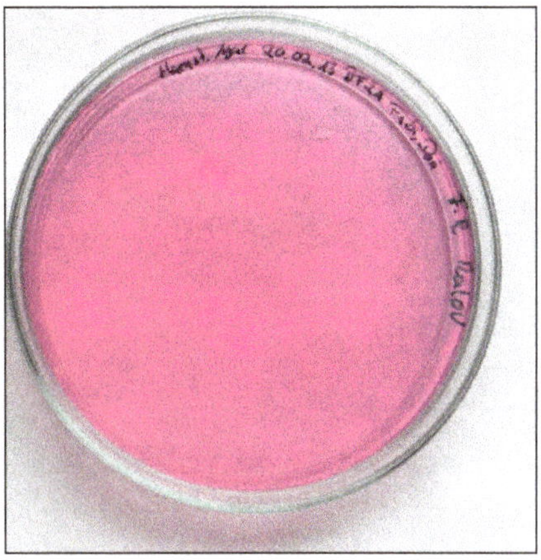

Raoultella planticola on urea agar.

The urease agar slant is used to measure an organism's ability to produce urease, an enzyme capable to digesting urea in carbon dioxide and ammonia through hydrolysis. Because ammonia is alkaline, the media contains phenol red, an indicator that changes from orange to pink when a pH increases above 8.1. When ammonia is increased to high enough concentrations, the media will change to a pink color, indicating the presence of urease production.

Voges–Proskauer Test

The Voges-Proskauer test detects whether a bacterium is producing the product acetoin from the digestion of glucose.

Cellular Fatty Acid based Identification

Mycolic Acid Analysis

Mycolic acid analysis has been an evolving field of study for gas-liquid chromatography, as it offers a solution to slow growth rates in *Mycobacterium*. Mycolic acid is a fatty acid found in the disease tuberculosis, offering a chemical target for diagnosticians to look for.

Nucleic Acid Extraction Techniques

Cesium Chloride/Ethidium Bromide Density Gradient Centrifugation

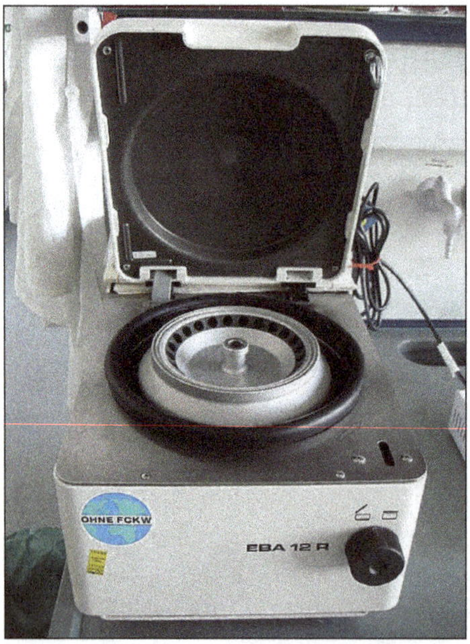

Tabletop centrifuge.

With high speed buoyant density ultracentrifugation, a density gradient is created with caesium chloride in water. DNA will go to the density that reflects its own, and ethidium bromide is then added to enhance the visuals the nucleic acid band provides.

Magnetic Bead Method

New extraction techniques have been developed using magnetic beads for the purification of nucleic acids by taking advantage of the charged and polymeric nature of long strand of DNA. Beads are both uncoated to increase surface are and yield, while others are more selective by being coated with functional groups that interact with the polymers present in microbes.

Phenol–Chloroform Extraction

Phenol-Chloroform extraction is a liquid-liquid method used by biochemists to separate nucleic acids from proteins and lipids after cells have been lysed.

Solid Phase Extraction

Solid phase extraction which separates long polymers like DNA from other substances found in the cells.

Methods with Electrophoretic Outputs

Gel electrophoresis is a technique to separate macromolecules by taking advantage of the charge on many of the molecules found in nucleic acids and protein.

Restriction Enzyme Based

Optical Mapping

Optical mapping is a technique using multiple restriction enzymes to create a genomic "barcode" which can be referenced back to diagnose an unknown microbe.

Pulsed-field Gel Electrophoresis

Pulsed-field gel electrophoresis is a technique used to separate large DNA in an electric field that periodically changes direction. By cutting segments of the DNA with restriction enzymes, pulse-field can be used to separate out the segments of DNA.

Restriction Enzymes then Gel Electrophoresis

Restriction enzymes are first used to recognize and then cut specific nucleic acid sequences. These cut pieces of DNA can be run through a gel electrophoresis to allow diagnostics of the organism by referencing back to previous gel electrophoresis results.

PCR-based

Multiple Loci VNTR Analysis

Multiple-locus VNTR analysis is a test used to test for variable number tandem repeats, which acts as a DNA fingerprint in microbial diagnostics.

Ribotyping

PCR Ribotyping is a rapid method for microbial diagnostics testing for rRNA in microbes.

DNA Sequence-based Method

Multi-locus Sequence Typing

Multilocus sequence typing is the sequencing of many single loci to diagnose an organism by referencing back to a database of known organisms.

Single-locus Sequence Typing

Single-locus sequence typing is the sequencing a single locus of an organism, about 1000 base pairs long, that can indicate the organism present.

Whole Genome Sequencing

Whole genome sequencing is used to diagnose an organism down to the individual base pairs.

AGRICULTURAL MICROBIOLOGY

Agricultural microbiology is a branch of microbiology dealing with plant-associated microbes and plant and animal diseases. It also deals with the microbiology of soil fertility, such as microbial degradation of organic matter and soil nutrient transformations. Done by Alpha B S Conteh of Njala University.

Soil Microorganisms

Bacteria:- more dominant group of microorganisms in the soil and equal to one half of the microbial biomass in soil. Population 100,000 to several hundred millions for gram of soil - Autochthnous - Zymogenous groups. Majority are Heterotrophs. (Common soil bacteria - Arthrobacter, Bacillus, Clostridium, Micrococcus).

Actinomycetes - intermediate group between bacteria and fungi. Numerous and widely distributed in soil. Abundance is next to bacteria. 104 - 108/g soil. 70% of soil actinomycetes are Streptomyces. Many of them are known to produce antibiotics. Population increases with depth of soil.

Fungi: More numerous in surface layers of well-aerated and cultivated soils-dominant in acid soils. Common genera in soil are Aspergillus, Mucor, Penicillium Trichoderma, Alternaria, Rhizopus. Algae – found in most of the soils in number ranges from 100 to 10,000 per g.

Protozoa: Unicellular – population ranges from 10,000 to 100,000 per g of soil. Most of the soil forms are flagellates, amoebae or ciliates. Derive their nutrition by devouring soil bacteria. Abundant in upper larger of the soil. They are regulating the biological equilibrium in soil.

Importance of Soil Microorganisms

- Involved in nutrient transformation process.
- Decomposition of resistant components of plant and animal tissue.
- Role in microbial antagonism.

SOIL MICROBIOLOGY

Soil microbiology is the study of microorganisms in soil, their functions, and how they affect soil properties. It is believed that between two and four billion years ago, the first ancient bacteria and microorganisms came about on Earth's oceans. These bacteria could fix nitrogen, in time multiplied, and as a result released oxygen into the atmosphere. This led to more advanced microorganisms, which are important because they affect soil structure and fertility. Soil microorganisms can be classified as bacteria, actinomycetes, fungi, algae and protozoa. Each of these groups has characteristics that define them and their functions in soil.

Up to 10 billion bacterial cells inhabit each gram of soil in and around plant roots, a region known as the rhizosphere. In 2011, a team detected more than 33,000 bacterial and archaeal species on sugar beet roots.

The composition of the rhizobiome can change rapidly in response to changes in the surrounding environment.

Bacteria

Bacteria and Archaea are the smallest organisms in soil apart from viruses. Bacteria and Archaea are prokaryotic. All of the *other* microorganisms are eukaryotic, which means they have a more advanced cell structure with internal organelles and the ability to reproduce sexually. A prokaryote has a very simple cell structure with no internal organelles. Bacteria and archaea are the most abundant microorganisms in the soil, and serve many important purposes, including nitrogen fixation.

Biochemical Processes

One of the most distinguished features of bacteria is their biochemical versatility . A bacterial genus called *Pseudomonas* can metabolize a wide range of chemicals and fertilizers. In contrast, another genus known as *Nitrobacter* can only derive its energy by turning nitrite into nitrate, which is also known as oxidation. The genus *Clostridium* is an example of bacterial versatility because it, unlike most species, can grow in the absence of oxygen, respiring anaerobically. Several species of *Pseudomonas*, such as *Pseudomonas aeruginosa* are able to respire both aerobically and anaerobically, using nitrate as the terminal electron acceptor.

Nitrogen Fixation

Bacteria are responsible for the process of nitrogen fixation, which is the conversion of atmospheric nitrogen into nitrogen-containing compounds (such as ammonia) that can be used by plants. Autotrophic bacteria derive their energy by making their own food through oxidation, like the *Nitrobacters* species, rather than feeding on plants or other organisms. These bacteria are responsible for nitrogen fixation. The amount of autotrophic bacteria is small compared to heterotrophic bacteria (the opposite of autotrophic bacteria, heterotrophic bacteria acquire energy by consuming plants or other microorganisms), but are very important because almost every plant and organism requires nitrogen in some way.

Actinomycetes

Actinomycetes are soil microorganisms. They are a type of bacteria, but they share some characteristics with fungi that are most likely a result of convergent evolution due to a common habitat and lifestyle.

Similarities to Fungi

Although they are members of the Bacteria kingdom, many actinomycetes share characteristics with fungi, including shape and branching properties, spore formation and secondary metabolite production.

- The mycelium branches in a manner similar to that of fungi.
- They form aerial mycelium as well as conidia.
- Their growth in liquid culture occurs as distinct clumps or pellets, rather than as a uniform turbid suspension as in bacteria.

Antibiotics

One of the most notable characteristics of the actinomycetes is their ability to produce antibiotics. Streptomycin, neomycin, erythromycin and tetracycline are only a few examples of these antibiotics. Streptomycin is used to treat tuberculosis and infections caused by certain bacteria and neomycin is used to reduce the risk of bacterial infection during surgery. Erythromycin is used to treat certain infections caused by bacteria, such as bronchitis, pertussis (whooping cough), pneumonia and ear, intestine, lung, urinary tract and skin infections.

Fungi

Fungi are abundant in soil, but bacteria are more abundant. Fungi are important in the soil as food sources for other, larger organisms, pathogens, beneficial symbiotic relationships with plants or other organisms and soil health. Fungi can be split into species based primarily on the size, shape and color of their reproductive spores, which are used to reproduce. Most of the environmental factors that influence the growth and distribution of bacteria and actinomycetes also influence fungi. The quality as well as quantity of organic matter in the soil has a direct correlation to the growth of fungi, because most fungi consume organic matter for nutrition. Fungi thrive in acidic environments, while bacteria and actinomycetes cannot survive in acid, which results in an abundance of fungi in acidic areas. Fungi also grow well in dry, arid soils because fungi are aerobic, or dependent on oxygen, and the higher the moisture content in the soil, the less oxygen is present for them.

Algae

Algae can make their own nutrients through photosynthesis. Photosynthesis converts light energy to chemical energy that can be stored as nutrients. For algae to grow, they must be exposed to light because photosynthesis requires light, so algae are typically distributed evenly wherever sunlight and moderate moisture is available. Algae do not have to be directly exposed to the Sun, but can live below the soil surface given uniform temperature and moisture conditions. Algae are also capable of performing nitrogen fixation.

Types

Algae can be split up into three main groups: the Cyanophyceae, the Chlorophyceae and the Bacillariaceae. The Cyanophyceae contain chlorophyll, which is the molecule that absorbs sunlight and uses that energy to make carbohydrates from carbon dioxide and water and also pigments that make it blue-green to violet in color. The Chlorophyceae usually only have chlorophyll in them which makes them green, and the Bacillariaceae contain chlorophyll as well as pigments that make the algae brown in color.

Blue-green Algae and Nitrogen Fixation

Blue-green algae, or Cyanophyceae, are responsible for nitrogen fixation. The amount of nitrogen they fix depends more on physiological and environmental factors rather than the organism's abilities. These factors include intensity of sunlight, concentration of inorganic and organic nitrogen sources and ambient temperature and stability.

Protozoa

Protozoa are eukaryotic organisms that were some of the first microorganisms to reproduce sexually, a significant evolutionary step from duplication of spores, like those that many other soil microorganisms depend on. Protozoa can be split up into three categories: flagellates, amoebae and ciliates.

Flagellates

Flagellates are the smallest members of the protozoa group, and can be divided further based on whether they can participate in photosynthesis. Nonchlorophyll-containing flagellates are not capable of photosynthesis because chlorophyll is the green pigment that absorbs sunlight. These flagellates are found mostly in soil. Flagellates that contain chlorophyll typically occur in aquatic conditions. Flagellates can be distinguished by their flagella, which is their means of movement. Some have several flagella, while other species only have one that resembles a long branch or appendage.

Amoebae

Amoebae are larger than flagellates and move in a different way. Amoebae can be distinguished from other protozoa by their slug-like properties and pseudopodia. A pseudopodium or "false foot" is a temporary obtrusion from the body of the amoeba that helps pull it along surfaces for movement or helps to pull in food. The amoeba does not have permanent appendages and the pseudopodium is more of a slime-like consistency than a flagellum.

Ciliates

Ciliates are the largest of the protozoa group, and move by means of short, numerous cilia that produce beating movements. Cilia resemble small, short hairs. They can move in different directions to move the organism, giving it more mobility than flagellates or amoebae.

Composition Regulation

Plant hormones salicylic acid, jasmonic acid and ethylene are key regulators of innate immunity in plant leaves. Mutants impaired in salicylic acid synthesis and signaling are hypersusceptible to microbes that colonize the host plant to obtain nutrients, whereas mutants impaired in jasmonic acid and ethylene synthesis and signaling are hypersusceptible to herbivorous insects and microbes that kill host cells to extract nutrients. The challenge of modulating a community of diverse microbes in plant roots is more involved than that of clearing a few pathogens from inside a plant leaf. Consequently, regulating root microbiome composition may require immune mechanisms other than those that control foliar microbes.

A 2015 study analyzed a panel of *Arabidopsis* hormone mutants impaired in synthesis or signaling of individual or combinations of plant hormones, the microbial community in the soil adjacent to the root and in bacteria living within root tissue. Changes in salicylic acid signaling stimulated a reproducible shift in the relative abundance of bacterial phyla in the endophytic compartment. These changes were consistent across many families within the affected phyla, indicating that salicylic acid may be a key regulator of microbiome community structure.

Classical plant defense hormones also function in plant growth, metabolism and abiotic stress responses, obscuring the precise mechanism by which salicylic acid regulates this microbiome.

During plant domestication, humans selected for traits related to plant improvement, but not for plant associations with a beneficial microbiome. Even minor changes in abundance of certain bacteria can have a major effect on plant defenses and physiology, with only minimal effects on overall microbiome structure.

Applications

Agriculture

Microbes can make nutrients and minerals in the soil available to plants, produce hormones that spur growth, stimulate the plant immune system and trigger or dampen stress responses. In general a more diverse soil microbiome results in fewer plant diseases and higher yield.

Farming can destroy soil's rhiziobiome (microbial ecosystem) by using soil amendments such as fertilizer and pesticide without compensating for their effects. By contrast, healthy soil can increase fertility in multiple ways, including supplying nutrients such as nitrogen and protecting against pests and disease, while reducing the need for water and other inputs. Some approaches may even allow agriculture in soils that were never considered viable.

The group of bacteria called rhizobia live inside the roots of legumes and fix nitrogen from the air into a biologically useful form.

Up to 30% of the carbon fixed by plants is excreted from the roots as so-called exudates—including sugars, amino acids, flavonoids, aliphatic acids, and fatty acids—that attract and feed beneficial microbial species while repelling and killing harmful ones.

FOOD MICROBIOLOGY

Food microbiology is the study of the microorganisms that inhibit, create, or contaminate food, including the study of microorganisms causing food spoilage, pathogens that may cause disease especially if food is improperly cooked or stored, those used to produce fermented foods such as cheese, yogurt, bread, beer, and wine, and those with other useful roles such as producing probiotics.

Food Safety

Food safety is a major focus of food microbiology. Numerous agents of disease, pathogens, are readily transmitted via food, including bacteria, and viruses. Microbial toxins are also possible contaminants of food. However, microorganisms and their products can also be used to combat these pathogenic microbes. Probiotic bacteria, including those that produce bacteriocins, can kill and inhibit pathogens. Alternatively, purified bacteriocins such as nisin can be added directly to food products. Finally, bacteriophages, viruses that only infect bacteria, can be used to kill bacterial pathogens. Thorough preparation of food, including proper cooking, eliminates most bacteria and viruses. However, *toxins produced* by contaminants may not be liable to change to non-toxic forms by heating or cooking the contaminated food due to other safety conditions.

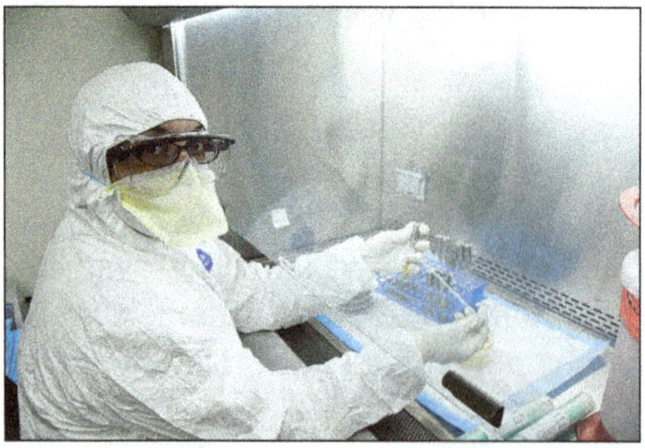

A microbiologist working in a biosafety laboratory tests for high risk pathogens in food.

Fermentation

Fermentation is one of the methods to preserve food and alter its quality. Yeast, especially *Saccharomyces cerevisiae*, is used to leaven bread, brew beer and make wine. Certain bacteria, including lactic acid bacteria, are used to make yogurt, cheese, hot sauce, pickles, fermented sausages and dishes such as kimchi. A common effect of these fermentations is that the food product is less hospitable to other microorganisms, including pathogens and spoilage-causing microorganisms, thus extending the food's shelf-life. Some cheese varieties also require molds to ripen and develop their characteristic flavors.

Microbial Biopolymers

Several microbially produced biopolymers are used in the food industry.

Alginate

Alginates can be used as thickening agents. Although listed here under the category 'Microbial polysaccharides', commercial alginates are currently only produced by extraction from brown seaweeds such as *Laminaria hyperborea* or *L. japonica*.

Poly-γ-glutamic Acid

Poly-γ-glutamic acid (γ-PGA) produced by various strains of *Bacillus* has potential applications as a thickener in the food industry.

Food Testing

To ensure safety of food products, microbiological tests such as testing for pathogens and spoilage organisms are required. This way the risk of contamination under normal use conditions can be examined and food poisoning outbreaks can be prevented. Testing of food products and ingredients is important along the whole supply chain as possible flaws of products can occur at every stage of production. Apart from detecting spoilage, microbiological tests can also determine germ content, identify yeasts and molds, and salmonella. For salmonella, scientists are also developing rapid and portable technologies capable of identifying unique variants of Salmonella.

Food microbiology laboratory at the Faculty of Food Technology, Latvia University of Life Sciences and Technologies.

Polymerase Chain Reaction (PCR) is a quick and inexpensive method to generate numbers of copies of a DNA fragment at a specific band ("PCR (Polymerase Chain Reaction)," 2008). For that reason, scientists are using PCR to detect different kinds of viruses or bacteria, such as HIV and anthrax based on their unique DNA patterns. Various kits are commercially available to help in food pathogen nucleic acids extraction, PCR detection, and differentiation. The detection of bacterial strands in food products is very important to everyone in the world, for it helps prevent the occurrence of food borne illness. Therefore, PCR is recognized as a DNA detector in order to amplify and trace the presence of pathogenic strands in different processed food.

INDUSTRIAL MICROBIOLOGY

Industrial microbiology is a branch of biotechnology that applies microbial sciences to create industrial products in mass quantities. There are multiple ways to manipulate a microorganism in order to increase maximum product yields. Introduction of mutations into an organism may be accomplished by introducing them to mutagens. Another way to increase production is by gene amplification, this is done by the use of plasmids, and vectors. The plasmids and/ or vectors are used to incorporate multiple copies of a specific gene that would allow more enzymes to be produced that eventually cause more product yield. The manipulation of organisms in order to yield a specific product has many applications to the real world like the production of some antibiotics, vitamins, enzymes, amino acids, solvents, alcohol and daily products. They can also be used in an agricultural application and use them as a biopesticede instead of using dangerous chemicals or as inoculants and help plant proliferation.

Medical Application

The medical application to industrial microbiology is the production of new drugs synthesized in a specific organism for medical purposes. Production of antibiotics is necessary for the treatment of many bacterial infections.Some natural occurring antibiotics and precursors, are produced through a process called fermentation. The microorganisms grow in a liquid media where the population size is controlled in order to yield the greatest amount of product. In this environment nutrient, pH, temperature, and oxygen are controlled also in order to maximize the amount of cells and cause them not to die before the production of the antibiotic of interest. Once the antibiotic is produced it must be extracted in order to yield an income.

Vitamins also get produced in massive quantities either by fermentation or biotransformation. Vitamin B 2 (riboflavin) for example is produced both ways. Biotransformation is mostly used for the production of riboflavin, and the carbon source starting material for this reaction is glucose. There are a few strains of microorganisms that were engineered to increase the yield of riboflavin produced. The most common organism used for this reaction is *Ashbya gossypii*. The fermentation process is another common way to produce riboflavin. The most common organism used for production of riboflavin through fermentation is *Eremothecium ashbyii*. Once riboflavin is produced it must be extracted from the broth, this is done by heating the cells for a certain amount of time, and then the cells can be filtered out of solution. Riboflavin is later purified and released as final product.

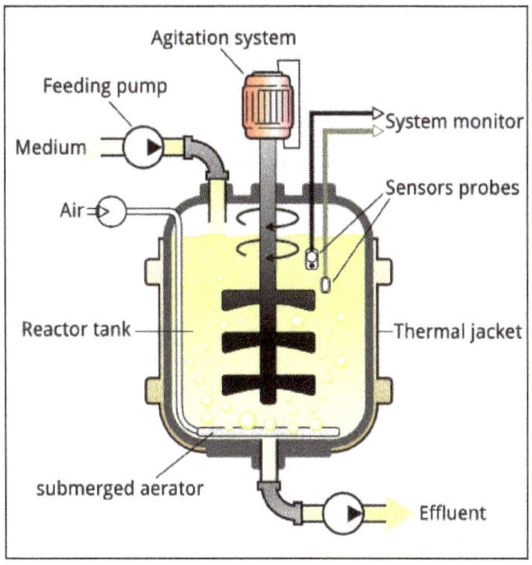

Representation of bioreactors. They are used to store the microbes while they are producing desired product of interest.

Enzymes can be produced through fermentation either by submerged fermentation and/ or by solid state fermentation. Submerged fermentation is referred to when the microorganisms are in contact with media. In this process the contact with oxygen is essential. The bioreactors/fermentors that are used to do these mass production of product can store up to 500 cubic meters in volume. Solid state fermentation is less common than submerged fermentation, but has many benefits. There is less need for the environment to be sterile since there is less water, there is a higher stability and concentration for the end product. Insulin synthesis is done through the fermentation process and the use of recombinant *E.coli* or yeast in order to make human insulin also called Humulin.

Food Industry Application

Fermentation

Fermentation is a reaction where sugar can be converted into a gas, alcohols or acids. Microorganisms like yeast and bacteria are used to massively produce the many things. Drinking alcohol also known as ethanol is produced by yeast and bacteria. Ethanol can also be used as a fuel source. The drinking alcohol is produced from natural sugars like glucose. Carbon dioxide is produced as a side product in this reaction and can be used to make bread, and can also be used to carbonate beverages. FermentationWine: Alcoholic beverages like beer and wine are fermented by microorganisms when there is no oxygen present.

In this process, once there is enough alcohol and carbon dioxide around in the media the yeast start to die due to the environment becoming toxic to them. There are many strains of yeast and bacteria that can tolerate different amounts of alcohol around in their environment before it becoming toxic, thus one can obtain different alcohol levels in beer and wine, just by selecting a different microbial strain.

Most yeast can tolerate between 10 and 15 percent alcohol, but there are some strains that can tolerate up to 21 percent alcohol. Dairy products like cheese and yogurt can also be made through fermentation using microbes. Cheese was produced as a way to preserve the nutrients obtained from milk, through fermentation thus elongating the shelf-life of the product. Microbes are used to convert the lactose sugars into lactic acid through fermentation. The bacteria used for such fermentation are usually from *Lactococci*, *Lactobacilli*, or *Streptococci* families. Sometimes these microbes are added before or after the acidification step needed for cheese production. Also these microbes are responsible for the different flavors of cheese, since they have enzymes that break-down milk sugars and fats into multiple building blocks.

Some other microbes like mold may be purposely introduced during or before the aging of the cheese, in order to give it a different flavor.

The production of yogurt starts from the pasteurization of milk, where undesired microbes are reduced or eliminated. Once the milk is pasteurized the milk is ready to be processed to reduce fat and liquid content, so what remains is mostly solid content. This can be done by drying the milk so that the liquid evaporates or by adding concentrated milk. Increasing the solid content of the milk also increases the nutritional value since the nutrients are more concentrated. After this step is accomplished, the milk is ready for fermentation where the milk gets inoculated with bacteria in hygienic stainless steel containers and then gets carefully monitored for lactic acid production, temperature and pH.

Agriculture Application

Biopesticide is a pesticide derivatized from a living organism or natural occurring substances. Biochemical pesticides can also be produced from naturally occurring substances that can control pest populations in a non-toxic matter. An example of a biochemical pesticide is garlic and pepper based insecticides, these work by repelling insects from the desired location. Microbial pesticides, usually a virus, bacterium, or fungus are used to control pest populations in a more specific manner. The most commonly used microbe for the production of microbial bio-pesticides is *Bacillus thuringiensis*, also known as Bt. This spore forming bacterium produces a delta-endotoxins in which it causes the insect or pest to stop feeding on the crop or plant because the endotoxin destroys the lining of the digestive system. Another mechanism that is used to reduce plant pathogens is by introducing other microbes that are non-pathogenic but compete for the rhizosphere, and succeed by producing anti fungal chemicals yielding plant growth.

Microbial inoculants are addition of microbes into a plant that would essentially help the plant grow by introducing nutrients, and stimulating plant growth. The preparation in mass quantities of any inoculum is performed by a process called fermentation. The first step to making it is by selecting a microbial strain, and letting it grow and increase in bacterial concentration. The greater the bacterial concentration the greater the fermentation yield would be (fermentation yield is ratio of bacterial concentration to mass of substrate) The bacterial concentration can be measured by monitoring the turbidity, wet or dry weight, or residual nutrient concentration. After the inoculant is ready then it gets transferred in a fermentor where oxygen and temperature most be highly monitored for the survival of the microbes and it can vary depending on the microbe that is being used. After enough time has passed for the microbes to be properly incubated the product is ready to be extracted purified and packaged.

Chemical Application

Synthesis of amino acids and organic solvents can also be made using microbes. The synthesis of essential amino acids such as are L-Methionine, L-Lysine, L-Tryptophan and the non-essential amino acid L-Glutamic acid are used today mainly for feed, food, and pharmaceutical industries. The production of these amino acids is due to *Corynebacterium glutamicum* and fermentation. *C.glutamicum* was engineered to be able to produce L-lysine and L-Glutamic acid is large quantities. L-Glutamic acid had a high demand for production because this amino acid is used to produce Monosodium glutamate (MSG) a food flavoring agent. In 2012 the total production of L-Glutamic acid was 2.2 million tons and is produced using a submerged fermentation technique inoculated with *C.glutamicum*. L-Lysine was originally produced from diaminopimelic acid (DAP) by *E.coli*, but once the *C.glutamicum* was discovered for the production of L-Glutamic acid. This organism and other autotrophs were later modified to yield other amino acids such as lysine, aspartate, methionine, isoleucine and threonine. L-Lysine is used for the feeding of pigs and chicken, as well as to treat nutrient deficiency, increase energy in a patient, and sometimes used to treat viral infections. L-Tryptophan is also produced through fermentation and by *Corynebacterium* and *E.coli,* though the production is not as large as the rest of the amino acids it is still produced for pharmaceutical purposes since it can be converted and used to produce neurotransmitters.

The production of organic solvents like acetone, butanol, and isopropanol through fermentation was one of the very first things to be produced by using bacteria, since achieving the necessary chirality of the products is easily achieved by using living systems. Solvent fermentation uses a series of *Clostridia* bacterial species. Solvent fermentation at first was not as productive as it is used today. The amount of bacteria required to yield a product was high, and the actual yield of product was low. Later technological advances were discovered that allowed scientist to genetically alter these strains to achieve a higher yield for these solvents. These Clostridial strains were transformed to have extra gene copies of enzymes necessary for solvent production, as well as being more tolerant to higher concentrations of the solvent being produced, since these bacteria have a range of product in which they can survive in before the environment becomes toxic. Yielding more strains that can use other subtrates was also another way to increase the productivity of these bacteria.

References

- Bacteriology, cell-biology, science-and-technology: encyclopedia.com, Retrieved 6 January, 2020
- Barton, Larry L.; Northup, Diana E. (9 September 2011). Microbial Ecology. Wiley-Blackwell. Oxford: John Wiley & Sons. P. 22. ISBN 978-1-118-01582-7. Retrieved 25 May 2020
- What-is-Virology.aspx, health: news-medical.net, Retrieved 10 May, 2020
- Thomson, R. B.; Wilson, M. L.; Weinstein, M. P. (2010). "The Clinical Microbiology Laboratory Director in the United States Hospital Setting". Journal of Clinical Microbiology. 48 (10): 3465–3469. Doi:10.1128/JCM.01575-10. PMC 2953135. PMID 20739497
- Protozoology-definition-morphology-and-life-cycle, protozoan-cell, parasitology: biologydiscus-sion.com, Retrieved 8 March, 2020
- Pubchem. "Resazurin sodium salt | c12h6nnao4 - pubchem". Pubchem.ncbi.nlm.nih.gov. Retrieved 2017-04-03
- Parasitology: environmentalscience.org, Retrieved 9 April, 2020
- Mcdaniel, L. D.; Young, E.; Delaney, J.; Ruhnau, F.; Ritchie, K. B.; Paul, J. H. (2010). "High Frequency of Horizontal Gene Transfer in the Oceans". Science. 330 (6000): 50. Bibcode:2010Sci...330...50M. Doi:10.1126/science.1192243. PMID 20929803
- Microbial-genetics, anatomy-and-physiology, medicine: encyclopedia.com, Retrieved 11 June, 2020
- "World Health Statistics 2012". World Health Organization. Archived from the original on 2013-04-20. Retrieved 2013-04-25
- Krásný, Lukáš; Hynek, Radovan; Hochel, Igor (2013-11-01). "Identification of bacteria using mass spectrometry techniques". International Journal of Mass Spectrometry. 353: 67–79. Doi:10.1016/j.ijms.2013.04.016

Classification of Microorganisms

3 CHAPTER

Microorganisms are classified into various types based upon their cellular composition, morphology, mean of locomotion, and reproduction. Their types are bacteria, fungus, virus, protozoa, archaea and microalgae. This chapter has been carefully written to provide an easy understanding of these classifications of microorganisms.

A microorganism is a living thing that is too small to be seen with the naked eye. Examples of microorganisms include bacteria, archaea, algae, protozoa, and microscopic animals such as the dust mite.

These microorganisms have been often under-appreciated and under-studied. Indeed, until Anton von Leeuwenhoek invented the microscope, we did not know they existed! Until that time, it was thought that phenomena such as illness and food spoilage were caused by "vapors" or "spontaneous generation".

Antonie van Leeuwenhoek.

Leeuwenhoek's invention of the microscope soon led Louis Pasteur to realize that many diseases were caused by microorganisms and to the practice of pasteurization, which kills microorganisms and makes our food products safe to eat today.

Now, we know that microorganisms are responsible for many things that happen in the world around us.

Microorganisms are found virtually everywhere, except for environments that have been made artificially sterile by humans. Even these must be constantly sterilized and carefully protected, lest microorganisms be tracked in from the outside world.

Microorganisms live in water, in soil, and on the skin and in the digestive tracts of animals. This is why all living things must have immune systems while many microorganisms can be helpful to them, some can be harmful and cause disease.

Like all organisms, microorganisms play important roles in the ecosystems they inhabit. Here are a few of their roles.

Examples of Microorganisms

Streptococcus Bacteria

Streptococcus is a group of bacteria that causes illness in humans. As the name suggests, streptococcus bacteria is the cause of strep throat and can also cause scarlet fever and, rarely, skin and muscle infections.

Streptococcus is a good example of the "dangerous" type of microorganism. On this list, we'll discuss microorganisms that are both dangerous and helpful to humans.

Malaria Parasite (Plasmodium)

Malaria is the deadliest disease to humans on Earth today. Transmitted from host to host by mosquito bites, it causes serious symptoms such as fever, hemolytic anemia, and convulsions as the Plasmodium parasite reproduces inside the host's body.

It may surprise you to learn that the Plasmodium parasite is not a bacteria rather, it is a eukaryotic microorganism that reproduces sexually and undergoes a multi-stage life cycle.

Plasmodium demonstrates the diversity of microorganisms which may be bacterial, eukaryotic, or even multicellular.

Lactobacillus Rhamnosus

If Streptococcus is a "bad" bacterium, Lactobacillus is a "good" type of bacterium. Lactobacillus are bacteria that live in the guts of healthy people, and may help us fight off diseases such as the stomach flu.

Lactobacillus is found in many yogurts. Some people even take highly concentrated doses of Lactobacillus in the form of "probiotic" pills or capsules in hopes of staying healthy.

Cyanobacteria

Cyanobacteria was one of the first types of life to evolve on Earth. Its modern descendant continues to play important ecological roles today. Cyanobacteria can turn carbon dioxide into oxygen and it can turn unusable inorganic nitrogen into organic forms that can be used to make proteins and more.

Because cyanobacteria was one of the first organisms on the planet, it likely had to do those things for itself there were no other nitrogen-fixing organisms around to partner up with.

Cyanobacteria is another great example of a "good" microorganism to which humans owe a great deal.

ARCHAEA

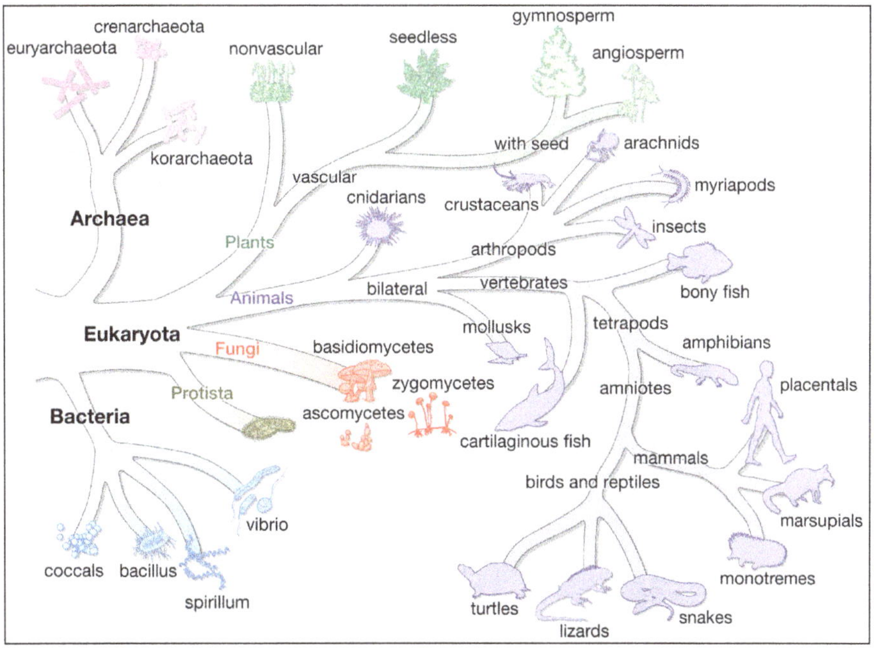

Life: Three domain classification.

Archaea is , any of a group of single-celled prokaryotic organisms (that is, organisms whose cells lack a defined nucleus) that have distinct molecular characteristics separating them from bacteria (the other, more prominent group of prokaryotes) as well as from eukaryotes (organisms, including plants and animals, whose cells contain a defined nucleus). Archaea is derived from the Greek word archaios, meaning "ancient" or "primitive," and indeed some archaea exhibit characteristics worthy of that name. Members of the archaea include: Pyrolobus fumarii, which holds the upper temperature limit for life at 113 °C (235 °F) and was found living in hydrothermal vents; species of Picrophilus, which were isolated from acidic soils in Japan and are the most acid-tolerant organisms known—capable of growth at around pH 0; and the methanogens, which produce methane gas as a metabolic by-product and are found in anaerobic environments, such as in marshes, hot springs, and the guts of animals, including humans. In some systems for classifying all of life, the archaea constitute one of three great domains of living creatures. In 1977 American microbiologist Carl Woese, on the basis of analyses of ribosomal RNA, proposed that the prokaryotes, long considered to be a single group of organisms (essentially, the bacteria), actually consist of two separate lineages. Woese called these two lineages the eubacteria and the archaebacteria. These names were subsequently changed to bacteria and archaea (the archaea being distinctly different from bacteria), but Woese's splitting of the prokaryotes into two groups has remained, and all living organisms are now considered by many biologists to fall into one of three great domains: Archaea, Bacteria, and Eukarya. Further molecular analysis has shown that domain Archaea consists of two major subdivisions, the Crenarchaeota and the Euryarchaeota, and one minor ancient lineage, the Korarchaeota. Other subdivisions have been proposed, including Nanoarchaeota and Thaumarchaeota.

Habitats of the Archaea

Archaea are microorganisms that define the limits of life on Earth. They were originally discovered and described in extreme environments, such as hydrothermal vents and terrestrial hot springs. They were also found in a diverse range of highly saline, acidic, and anaerobic environments.

Archaea.

Although many of the cultured archaea are extremophiles, these organisms in their respective extreme habitats represent only a minority of the total diversity of the Archaea domain. The majority of archaea cannot be cultured within the laboratory setting, and their ubiquitous presence in global habitats has been realized through the use of culture-independent techniques. One commonly used culture-independent technique is the isolation and analysis of nucleic acids (i.e., DNA and RNA) directly from an environment, rather than the analysis of cultured samples isolated from the same environment. Culture-independent studies have shown that archaea are abundant and fulfill important ecological roles in cold and temperate ecosystems. Uncultivated organisms in the subdivision Crenarchaeota are postulated to be the most abundant ammonia-oxidizing organisms in soils and to account for a large proportion (roughly 20 percent) of the microorganisms present in the picoplankton in the world's oceans. In the subdivision Euryarchaeota, uncultivated organisms in deep-sea marine sediments are responsible for the removal of methane, a potent greenhouse gas, via anaerobic oxidation of methane stored in these sediments. In contrast, uncultivated methanogenic (methane-producing) euryarchaea from terrestrial anaerobic environments, such as rice fields, are estimated to generate approximately 10–25 percent of global methane emissions.

The cultured representatives of the Crenarchaeota are from high-temperature environments, such as hot springs and submarine hydrothermal vents. Likewise, cultured members of the Euryarchaeota include organisms isolated from hot environments, organisms that are methanogenic, and organisms that grow vigorously in high-salt environments (halophiles). Organisms in the Korarchaeota lineage and the proposed Nanoarchaeota lineage also inhabit high-temperature environments; however, the nanoarchaea are highly unusual because they grow and divide on the surface of another archaea, Ignicoccus. Nanoarchaea, which were discovered in 2002, contain both the smallest known living cell (1/100th the size of Escherichia coli) and the smallest known genome (490 kilobases [1 kilobase = 1,000 base pairs of DNA]; for comparison, the human genome contains 3,000,000 kilobases). Members of the Korarchaeota and Nanoarchaeota have not been detected in pure culture; rather, they have been detected only in mixed laboratory cultures.

Archaea are also found living in association with eukaryotes. For example, methanogenic archaea are present in the digestive systems of some animals, including humans. Some archaea form symbiotic relationships with sponges. In fact, Cenarchaeum symbiosum was grown in the laboratory with its host sponge and was the first nonthermophilic Crenarchaeota to be cultured and described. It was the first organism considered for classification in the proposed Thaumarchaeota lineage.

Characteristics of the Archaea

Although the domains Bacteria, Archaea, and Eukarya were founded on genetic criteria, biochemical properties also indicate that the archaea form an independent group within the prokaryotes and that they share traits with both the bacteria and the eukaryotes. Major examples of these traits include:

1. Cell walls: Virtually all bacteria contain peptidoglycan in their cell walls; however, archaea and eukaryotes lack peptidoglycan. Various types of cell walls exist in the archaea. Therefore, the absence or presence of peptidoglycan is a distin-guishing feature between the archaea and bacteria.

2. Fatty acids: Bacteria and eukaryotes produce membrane lipids consisting of fatty acids linked by ester bonds to a molecule of glycerol. In contrast, the archaea have ether bonds connecting fatty acids to molecules of glycerol.

Although a few bacteria also contain ether-linked lipids, no archaea have been discovered that contain ester-linked lipids.

3. Complexity of RNA polymerase: Transcription within all types of organisms is performed by an enzyme called RNA polymerase, which copies a DNA template into an RNA product. Bacteria contain a simple RNA polymerase consisting of four polypeptides. However, both archaea and eukaryotes have multiple RNA polymerases that contain multiple polypeptides. For example, the RNA polymerases of archaea contain more than eight polypeptides. The RNA polymerases of eukaryotes also consist of a high number of polypeptides (10–12), with the relative sizes of the polypeptides being similar to that of hyperthermophilic archaeal RNA polymerase. Therefore, the archaeal RNA polymerases more closely resemble RNA polymerases of eukaryotes rather than those of bacteria.
4. Protein synthesis: Various features of protein synthesis in the archaea are similar to those of eukaryotes but not of bacteria. A prominent difference is that bacteria have an initiator tRNA (transfer RNA) that has a modified methionine, whereas eukaryotes and archaea have an initiator tRNA with an unmodified methionine.
5. Metabolism: Various types of metabolism exist in both archaea and bacteria that do not exist in eukaryotes, including nitrogen fixation, denitrification, chemolithotrophy, and hyperthermophilic growth. Methanogenesis (the production of methane as a metabolic by-product) occurs only in the domain Archaea, specifically in the subdivision Euryarchaeota. Classical photosynthesis using chloro-phyll has not been found in any archaea.

The metabolic strategies utilized by the archaea are thought to be extraordinarily diverse in nature. For example, halophilic archaea appear to be able to thrive in high-salt environments because they house a special set of genes encoding enzymes for a metabolic pathway that limits osmosis. That metabolic pathway, known as the methylaspartate pathway, represents a unique type of anaplerosis (the process of replenishing supplies of metabolic intermediates; in this instance the intermediate is methylaspartate). Halophilic archaeans, which include Haloarcula marismortui, a model organism used in scientific research, are thought to have acquired the unique set of genes for the methylaspartate pathway via a process known as horizontal gene transfer, in which genes are passed from one species to another.

BACTERIA

Bacteria is any of a group of microscopic single-celled organisms that live in enormous numbers in almost every environment on Earth, from deep-sea vents to deep below Earth's surface to the digestive tracts of humans.

Bacteria lack a membrane-bound nucleus and other internal structures and are therefore ranked among the unicellular life-forms called prokaryotes. Prokaryotes are the dominant living creatures on Earth, having been present for perhaps three-quarters of Earth history and having adapted to almost all available ecological habitats. As a group, they display exceedingly diverse metabolic capabilities and can use almost any organic compound, and some inorganic compounds, as a food source. Some bacteria can cause diseases in humans, animals, or plants, but most are harmless and are beneficial ecological agents whose metabolic activities sustain higher life-forms. Other bacteria are symbionts of plants and invertebrates, where they carry out important functions for the host, such as nitrogen fixation and cellulose degradation. Without prokaryotes, soil would not be fertile, and dead organic material would decay much more slowly. Some bacteria are widely used in the preparation of foods, chemicals, and antibiotics. Studies of the relationships between different groups of bacteria continue to yield new insights into the origin of life on Earth and mechanisms of evolution.

The Bacterial Cell

Bacteria as Prokaryotes

All living organisms on Earth are made up of one of two basic types of cells: eukaryotic cells, in which the genetic material is enclosed within a nuclear membrane, or prokaryotic cells, in which the genetic material is not separated from the rest of the cell. Traditionally, all prokaryotic cells were called bacteria and were classified in the prokaryotic kingdom Monera. However, their classification as Monera, equivalent in taxonomy to the other kingdoms—Plantae, Animalia, Fungi, and Protista—understated the remarkable genetic and metabolic diversity exhibited by prokaryotic cells relative to eukaryotic cells. In the late 1970s American microbiologist Carl Woese pioneered a major change in classification by placing all organisms into three domains—Eukarya, Bacteria (originally called Eubacteria), and Archaea (originally called Archaebacteria)—to reflect the three ancient lines of evolution. The prokaryotic organisms that were formerly known as bacteria were then divided into two of these domains, Bacteria and Archaea. Bacteria and Archaea are superficially similar; for example, they do not have intracellular organelles, and they have circular DNA. However, they are fundamentally distinct, and their separation is based on the genetic evidence for their ancient and separate evolutionary lineages, as well as fundamental differences in their chemistry and physiology. Members of these two prokaryotic domains are as different from one another as they are from eukaryotic cells.

Bacterial cells differ from animal cells and plant cells in several ways. One fundamental difference is that bacterial cells lack intracellular organelles, such as mitochondria, chloroplasts, and a nucleus, which are present in both animal cells and plant cells.

Bacterial cells differ from animal cells and plant cells in several ways. One fundamental difference is that bacterial cells lack intracellular organelles, such as mitochondria, chloroplasts, and a nucleus, which are present in both animal cells and plant cells.

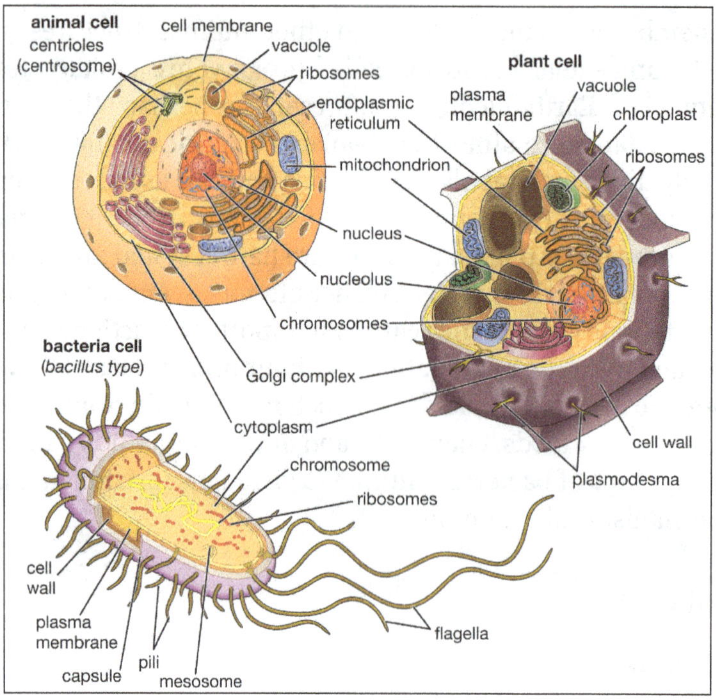

Some typical cells.

Prokaryotic cells (i.e., Bacteria and Archaea) are fundamentally different from the eukaryotic cells that constitute other forms of life. Prokaryotic cells are defined by a much simpler design than is found in eukaryotic cells. The most-apparent simplification is the lack of intracellular organelles, which are features characteristic of eukaryotic cells. Organelles are discrete membrane-enclosed structures that are contained in the cytoplasm and include the nucleus, where genetic information is retained, copied, and expressed; the mitochondria and chloroplasts, where chemical or light energy is converted into metabolic energy; the lysosome, where ingested proteins are digested and other nutrients are made available; and the endoplasmic reticulum and the Golgi apparatus, where the proteins that are synthesized by and released from the cell are assembled, modified, and exported. All of the activities performed by organelles also take place in bacteria, but they are not carried out by specialized structures. In addition, prokaryotic cells are usually much smaller than eukaryotic cells. The small size, simple design, and broad metabolic capabilities of bacteria allow them to grow and divide very rapidly and to inhabit and flourish in almost any environment.

Prokaryotic and eukaryotic cells differ in many other ways, including lipid composition, structure of key metabolic enzymes, responses to antibiotics and toxins, and the mechanism of expression of genetic information. Eukaryotic organisms contain multiple linear chromosomes with genes that are much larger than they need to be to encode the synthesis of proteins. Substantial portions of the ribonucleic acid (RNA) copy of the genetic information (deoxyribonucleic acid, or DNA) are discarded, and the remaining messenger RNA (mRNA) is substantially modified before it is translated into protein.

In contrast, bacteria have one circular chromosome that contains all of their genetic information, and their mRNAs are exact copies of their genes and are not modified.

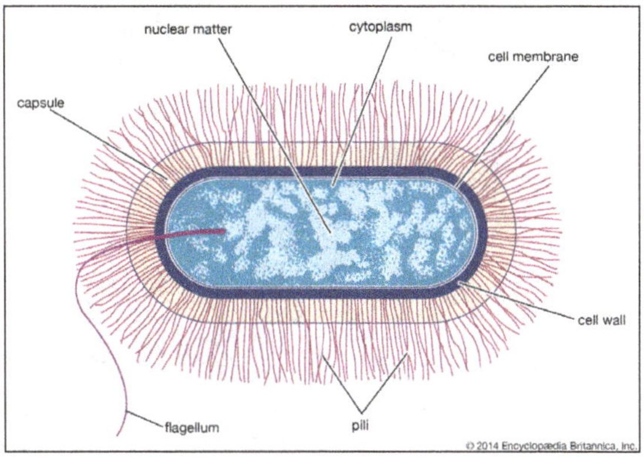

Structure of typical bacterial cell of the bacillus type.

Diversity of Structure of Bacteria

Although bacterial cells are much smaller and simpler in structure than eukaryotic cells, the bacteria are an exceedingly diverse group of organisms that differ in size, shape, habitat, and metabolism. Much of the knowledge about bacteria has come from studies of disease-causing bacteria, which are more readily isolated in pure culture and more easily investigated than are many of the free-living species of bacteria. It must be noted that many free-living bacteria are quite different from the bacteria that are adapted to live as animal parasites or symbionts. Thus, there are no absolute rules about bacterial composition or structure, and there are many exceptions to any general statement.

Individual bacteria can assume one of three basic shapes: spherical (coccus), rodlike (bacillus), or curved (vibrio, spirillum, or spirochete). Considerable variation is seen in the actual shapes of bacteria, and cells can be stretched or compressed in one dimension. Bacteria that do not separate from one another after cell division form characteristic clusters that are helpful in their identification. For example, some cocci are found mainly in pairs, including Streptococcus pneumoniae, a pneumococcus that causes bacterial lobar pneumonia, and Neisseria gonorrhoeae, a gonococcus that causes the sexually transmitted disease gonorrhea. Most streptococci resemble a long strand of beads, whereas the staphylococci form random clumps (the name "staphylococci" is derived from the Greek word staphyle, meaning "cluster of grapes"). In addition, some coccal bacteria occur as square or cubical packets. The rod-shaped bacilli usually occur singly, but some strains form long chains, such as rods of the corynebacteria, normal inhabitants of the mouth that are frequently attached to one another at random angles. Some bacilli have pointed ends, whereas others have squared ends, and some rods are bent into a comma shape. These bent rods are often called vibrios and include Vibrio cholerae, which causes cholera. Other shapes of bacteria include the spirilla, which are bent and rebent, and the spirochetes, which form a helix similar to a corkscrew, in which the cell body is wrapped around a central fibre called the axial filament.

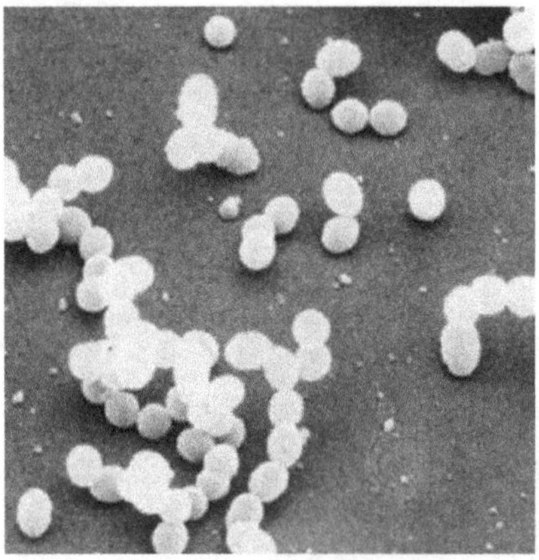

The bacterium *Streptococcus mutans* is an example of a spherical (coccus) bacterium. This species of bacteria commonly aggregates into pairs and short chains.

Bacteria are the smallest living entities. An average-size bacterium—such as the rod-shaped Escherichia coli, a normal inhabitant of the intestinal tract of humans and animals—is about 2 micrometres (μm; millionths of a metre) long and 0.5 μm in diameter, and the spherical cells of Staphylococcus aureus are up to 1 μm in diameter. A few bacterial types are even smaller, such as Mycoplasma pneumoniae, which is one of the smallest bacteria, ranging from about 0.1 to 0.25 μm in width and roughly 1 to 1.5 μm in length; the rod-shaped Bordetella pertussis, which is the causative agent of whooping cough, ranging from 0.2 to 0.5 μm in diameter and 0.5 to 1 μm in length; and the corkscrew-shaped Treponema pallidum, which is the causative agent of syphilis, averaging only 0.1 to 0.2 μm in diameter but 6 to 15 μm in length. The cyanobacterium Synechococcus averages about 0.5 to 1.6 μm in diameter. Some bacteria are relatively large, such as Azotobacter, which has diameters of 2 to 5 μm or more; and Achromatium, which has a minimum width of 5 μm and a maximum length of 100 μm, depending on the species. Giant bacteria can be visible with the unaided eye, such as Thiomargarita namibiensis, which averages 750 μm in diameter, and the rod-shaped Epulopiscium fishelsoni, which ranges from 30 to more than 600 μm in length.

Bacteria are unicellular microorganisms and thus are generally not organized into tissues. Each bacterium grows and divides independently of any other bacterium, although aggregates of bacteria, sometimes containing members of different species, are frequently found. Many bacteria can form aggregated structures called biofilms. Organisms in biofilms often display substantially different properties from the same organism in the individual state or the planktonic state. Bacteria that have aggregated into biofilms can communicate information about population size and metabolic state. This type of communication is called quorum sensing and operates by the production of small molecules called autoinducers or pheromones. The concentration of quorum-sensing molecules—most commonly peptides or acylated homoserine lactones (AHLs; special signaling chemicals)—is related to the number of bacteria of the same or different species that are in the biofilm and helps coordinate the behaviour of the biofilm.

Morphological Features of Bacteria

The Gram Stain

Bacteria are so small that their presence was only first recognized in 1677, when the Dutch naturalist Antonie van Leeuwenhoek saw microscopic organisms in a variety of substances with the aid of primitive microscopes (more similar in design to modern magnifying glasses than modern microscopes), some of which were capable of more than 200-fold magnification. Now bacteria are usually examined under light microscopes capable of more than 1,000-fold magnification; however, details of their internal structure can be observed only with the aid of much more powerful transmission electron microscopes. Unless special phase-contrast microscopes are used, bacteria have to be stained with a coloured dye so that they will stand out from their background.

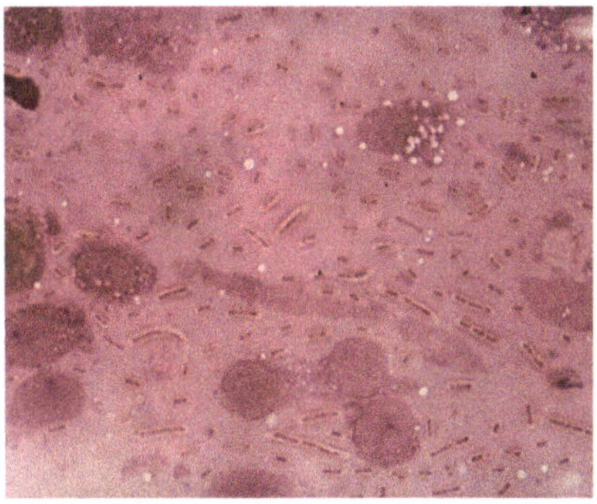

Gram-negative bacilli, *Klebsiella pneumoniae*, isolated from a lung abscess in a patient with pneumonia.

One of the most useful staining reactions for bacteria is called the Gram stain, developed in 1884 by the Danish physician Hans Christian Gram. Bacteria in suspension are fixed to a glass slide by brief heating and then exposed to two dyes that combine to form a large blue dye complex within each cell. When the slide is flushed with an alcohol solution, gram-positive bacteria retain the blue colour and gram-negative bacteria lose the blue colour. The slide is then stained with a weaker pink dye that causes the gram-negative bacteria to become pink, whereas the gram-positive bacteria remain blue. The Gram stain reacts to differences in the structure of the bacterial cell surface, differences that are apparent when the cells are viewed under an electron microscope.

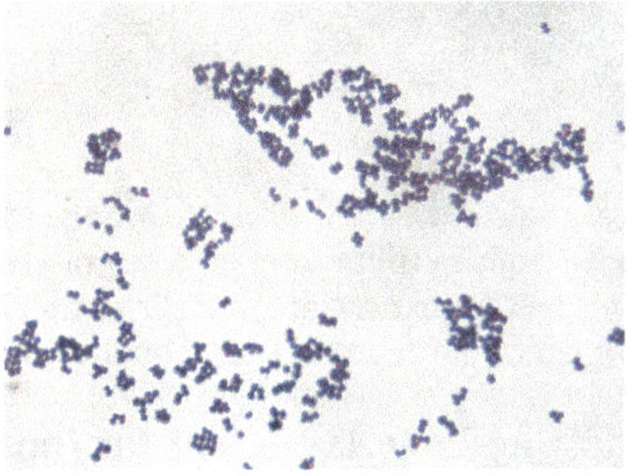

Gram-positive cocci, *Staphylococcus aureus*, in a laboratory culture.

The Cell Envelope

The bacterial cell surface (or envelope) can vary considerably in its structure, and it plays a central role in the properties and capabilities of the cell. The one feature present in all cells is the cytoplasmic membrane, which separates the inside of the cell from its external environment, regulates the flow of nutrients, maintains the proper intracellular milieu, and prevents the loss of the cell's contents. The cytoplasmic membrane carries out many necessary cellular functions, including energy generation, protein secretion, chromosome segregation, and efficient active transport of nutrients. It is a typical unit membrane composed of proteins and lipids, basically similar to the membrane that surrounds all eukaryotic cells. It appears in electron micrographs as a triple-layered structure of lipids and proteins that completely surround the cytoplasm.

Peptidoglycan layer of Bacillus coagulans A portion of the gram-positive bacterium Bacillus coagulans showing the cell wall's thick peptidoglycan layer that surrounds the cell membrane.

Lying outside of this membrane is a rigid wall that determines the shape of the bacterial cell. The wall is made of a huge molecule called peptidoglycan (or murein). In gram-positive bacteria the peptidoglycan forms a thick meshlike layer that retains the blue dye of the Gram stain by trapping it in the cell. In contrast, in gram-negative bacteria the peptidoglycan layer is very thin (only one or two molecules deep), and the blue dye is easily washed out of the cell.

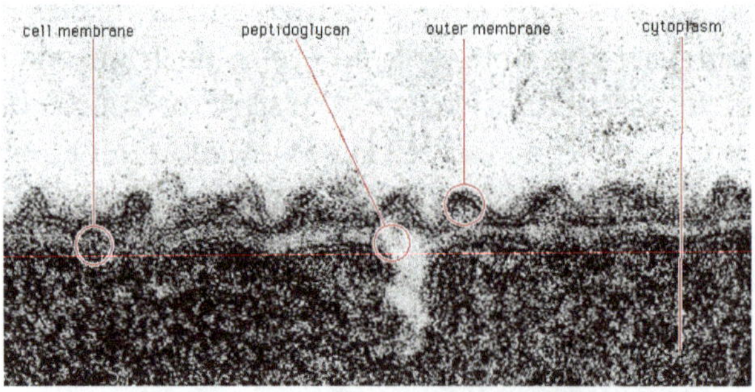

Peptidoglycan layer of Aquaspirillum serpens The gram-negative bacterium Aquaspirillum serpens has a thin peptidoglycan layer that lies between the cell membrane and the outer membrane.

Peptidoglycan occurs only in the Bacteria (except for those without a cell wall, such as Mycoplasma). Peptidoglycan is a long-chain polymer of two repeating sugars

(N-acetylglucosamine and N-acetyl muramic acid), in which adjacent sugar chains are linked to one another by peptide bridges that confer rigid stability. The nature of the peptide bridges differs considerably between species of bacteria but in general consists of four amino acids: L-alanine linked to D-glutamic acid, linked to either diaminopimelic acid in gram-negative bacteria or L-lysine, L-ornithine, or diaminopimelic acid in gram-positive bacteria, which is finally linked to D-alanine. In gram-negative bacteria the peptide bridges connect the D-alanine on one chain to the diaminopimelic acid on another chain. In gram-positive bacteria there can be an additional peptide chain that extends the reach of the cross-link; for example, there is an additional bridge of five glycines in Staphylococcus aureus.

Peptidoglycan synthesis is the target of many useful antimicrobial agents, including the β-lactam antibiotics (e.g., penicillin) that block the cross-linking of the peptide bridges. Some of the proteins that animals synthesize as natural antibacterial defense factors attack the cell walls of bacteria. For example, an enzyme called lysozyme splits the sugar chains that are the backbone of peptidoglycan molecules. The action of any of these agents weakens the cell wall and disrupts the bacterium.

In gram-positive bacteria the cell wall is composed mainly of a thick peptidoglycan meshwork interwoven with other polymers called teichoic acids and some proteins or lipids. In contrast, gram-negative bacteria have a complex cell wall that is composed of multiple layers in which an outer membrane layer lies on top of a thin peptidoglycan layer. This outer membrane is composed of phospholipids, which are complex lipids that contain molecules of phosphate, and lipopolysaccharides, which are complex lipids that are anchored in the outer membrane of cells by their lipid end and have a long chain of sugars extending away from the cell into the medium. Lipopolysaccharides, often called endotoxins, are toxic to animals and humans; their presence in the bloodstream can cause fever, shock, and even death. For most gram-negative bacteria, the outer membrane forms a barrier to the passage of many chemicals that would be harmful to the bacterium, such as dyes and detergents that normally dissolve cellular membranes. Impermeability to oil-soluble compounds is not seen in other biological membranes and results from the presence of lipopolysaccharides in the membrane and from the unusual character of the outer membrane proteins. As evidence of the ability of the outer membrane to confer resistance to harsh environmental conditions, some gram-negative bacteria grow well in oil slicks, jet fuel tanks, acid mine drainage, and even bottles of disinfectants.

The Archaea have markedly different surface structures from the Bacteria. They do not have peptidoglycan; instead, their membrane lipids are made up of branched isoprenoids linked to glycerol by ether bonds. Some archaea have a wall material that is similar to peptidoglycan, except that the specific sugar linked to the amino acid bridges is not muramic acid but talosaminuronic acid. Many other archaeal species use proteins as the basic constituent of their walls, and some lack a rigid wall.

Capsules and Slime Layers

Many bacterial cells secrete some extracellular material in the form of a capsule or a slime layer. A slime layer is loosely associated with the bacterium and can be easily washed off, whereas a capsule is attached tightly to the bacterium and has definite boundaries. Capsules can be seen under a light microscope by placing the cells in a suspension of India ink. The capsules exclude the ink and appear as clear halos surrounding the bacterial cells. Capsules are usually polymers of simple sugars (polysaccharides), although the capsule of Bacillus anthracis is made of polyglutamic acid. Most capsules are hydrophilic ("water-loving") and may help the bacterium avoid desiccation (dehydration) by preventing water loss. Capsules can protect a bacterial cell from ingestion and destruction by white blood cells (phagocytosis). While the exact mechanism for escaping phagocytosis is unclear, it may occur because capsules make bacterial surface components more slippery, helping the bacterium to escape engulfment by phagocytic cells. The presence of a capsule in Streptococcus pneumoniae is the most important factor in its ability to cause pneumonia. Mutant strains of S. pneumoniae that have lost the ability to form a capsule are readily taken up by white blood cells and do not cause disease. The association of virulence and capsule formation is also found in many other species of bacteria.

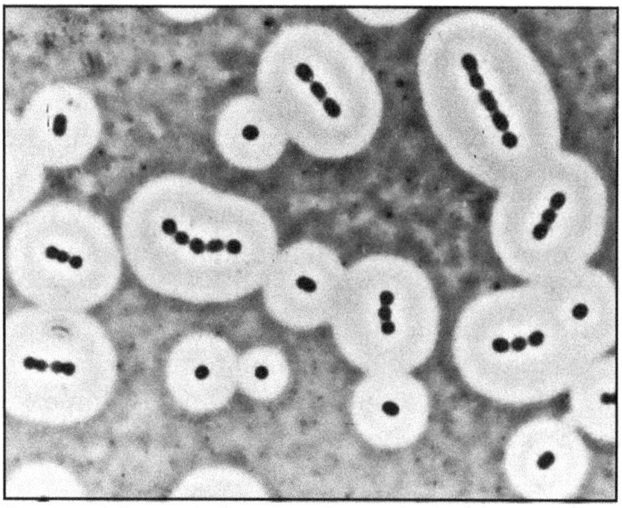

The capsular material surrounding these bacteria (Acinetobacter calcoaceticus) is revealed in a suspension of India ink and viewed through a light microscope (magnified about 2,500×).

The capsular material surrounding these bacteria (Acinetobacter calcoaceticus) is revealed in a suspension of India ink and viewed through a light microscope (magnified about 2,500×).

A capsular layer of extracellular polysaccharide material can enclose many bacteria into a biofilm and serves many functions. Streptococcus mutans, which causes dental caries, splits the sucrose in food and uses one of the sugars to build its capsule, which sticks tightly to the tooth. The bacteria that are trapped in the capsule use the other sugar to fuel their metabolism and produce a strong acid (lactic acid) that attacks the tooth enamel. When Pseudomonas aeruginosa colonizes the lungs of persons with cystic fibrosis, it produces a thick capsular polymer of alginic acid that contributes to the difficulty of eradicating the bacterium. Bacteria of the genus Zoogloea secrete fibres of cellulose that enmesh the bacteria into a floc that floats on the surface of liquid and keeps the bacteria exposed to air, a requirement for the metabolism of this genus. A few rod-shaped bacteria, such as Sphaerotilus, secrete long chemically complex tubular sheaths that enclose substantial numbers of the bacteria. The sheaths of these and many other environmental bacteria can become encrusted with iron or manganese oxides.

Flagella, Fimbriae and Pili

Many bacteria are motile, able to swim through a liquid medium or glide or swarm across a solid surface. Swimming and swarming bacteria possess flagella, which are the extracellular appendages needed for motility. Flagella are long, helical filaments made of a single type of protein and located either at the ends of rod-shaped cells, as in Vibrio cholerae or Pseudomonas aeruginosa, or all over the cell surface, as in Escherichia coli. Flagella can be found on both gram-positive and gram-negative rods but are rare on cocci and are trapped in the axial filament in the spirochetes. The flagellum is attached at its base to a basal body in the cell membrane. The protomotive force generated at the membrane is used to turn the flagellar filament, in the manner of a turbine driven by the flow of hydrogen ions through the basal body into the cell. When the flagella are rotating in a counterclockwise direction, the bacterial cell swims in a straight line; clockwise rotation results in swimming in the opposite direction or, if there is more than one flagellum per cell, in random tumbling. Chemotaxis allows a bacterium to adjust its swimming behaviour so that it can sense and migrate toward increasing levels of an attractant chemical or away from a repellent one.

Not only are bacteria able to swim or glide toward more favourable environments, but they also have appendages that allow them to adhere to surfaces and keep from being washed away by flowing fluids. Some bacteria, such as E. coli and Neisseria gonorrhoeae, produce straight, rigid, spikelike projections called fimbriae (Latin for "threads" or "fibres") or pili (Latin for "hairs"), which extend from the surface of the bacterium and attach to specific sugars on other cells—for these strains, intestinal or urinary-tract epithelial cells, respectively. Fimbriae are present only in gram-negative bacteria. Certain pili (called sex pili) are used to allow one bacterium to recognize and adhere to another in a process of sexual mating called conjugation (see below Bacterial reproduction). Many aquatic bacteria produce an acidic mucopolysaccharide holdfast, which allows them to adhere tightly to rocks or other surfaces.

The Cytoplasm

Although bacteria differ substantially in their surface structures, their interior contents are quite similar and display relatively few structural features.

Genetic Content

The genetic information of all cells resides in the sequence of nitrogenous bases in the extremely long molecules of DNA. Unlike the DNA in eukaryotic cells, which resides in the nucleus, DNA in bacterial cells is not sequestered in a membrane-bound organelle but appears as a long coil distributed through the cytoplasm. In many bacteria the DNA is present as a single circular chromosome, although some bacteria may contain two chromosomes, and in some cases the DNA is linear rather than circular. A variable number of smaller, usually circular (though sometimes linear) DNA molecules, called plasmids, can carry auxiliary information.

The sequence of bases in the DNA has been determined for hundreds of bacteria. The amount of DNA in bacterial chromosomes ranges from 580,000 base pairs in Mycoplasma genitalium to 4,700,000 base pairs in E. coli to roughly 9,450,000 base pairs in Myxococcus xanthus. Sorangium cellulosum, a myxobacterium, has one of the largest bacterial genomes, containing in excess of 13 million base pairs. The length of the E. coli chromosome, if removed from the cell and stretched to its fullest extent, is about 1.2 mm, which is striking in view of the fact that the length of the cell is about 0.001 mm.

As in all organisms, bacterial DNA contains the four nitrogenous bases adenine (A), cytosine (C), guanine (G), and thymine (T). The rules of base pairing for double-stranded DNA molecules require that the number of adenine and thymine bases be equal and that the number of cytosine and guanine bases also be equal. The relationship between the number of pairs of G and C bases and the number of pairs of A and T bases is an important indicator of evolutionary and adaptive genetic changes within an organism. The proportion, or molar ratio, of G + C can be measured as G + C divided by the sum of all the bases (A + T + G + C) multiplied by 100 percent. The extent to which G + C ratios vary between organisms may be considerable. In plants and animals, the proportion of G + C is about 50 percent. A far wider range in the proportion of G + C is seen in prokaryotes, extending from about 25 percent in most Mycoplasma to about 50 percent in E. coli to nearly 75 percent in Micrococcus, actinomycetes, and fruiting myxobacteria. The G + C content within a species in a single genus, however, is very similar.

Cytoplasmic Structures

The cytoplasm of bacteria contains high concentrations of enzymes, metabolites, and salts. In addition, the proteins of the cell are made on ribosomes that are scattered throughout the cytoplasm. Bacterial ribosomes are different from ribosomes in eukaryotic cells in that they are smaller, have fewer constituents (consist of three types of ribosomal RNA [rRNA] and 55 proteins, as opposed to four types of rRNA and 78 proteins in eukaryotes), and are inhibited by different antibiotics than those that act on eukaryotic ribosomes.

There are numerous inclusion bodies, or granules, in the bacterial cytoplasm. These bodies are never enclosed by a membrane and serve as storage vessels. Glycogen, which is a polymer of glucose, is stored as a reserve of carbohydrate and energy. Volutin, or metachromatic granules, contains polymerized phosphate and represents a storage form for inorganic phosphate and energy. Many bacteria possess lipid droplets that contain polymeric esters of poly-β-hydroxybutyric acid or related compounds. This is in contrast to eukaryotes, which use lipid droplets to store triglycerides. In bacteria, storage granules are produced under favourable growth conditions and are consumed after the nutrients have been depleted from the medium. Many aquatic bacteria produce gas vacuoles, which are protein-bound structures that contain air and allow the bacteria to adjust their buoyancy. Bacteria can also have internal membranous structures that form as outgrowths of the cytoplasmic membrane.

Biotypes of Bacteria

The fact that pathogenic bacteria are constantly battling their host's immune system might account for the bewildering number of different strains, or types, of bacteria that belong to the same species but are distinguishable by serological tests. Microbiologists often identify bacteria by the presence of specific molecules on their cell surfaces, which are detected with specific antibodies. Antibodies are serum proteins that bind very tightly to foreign molecules (antigens) in an immune reaction aimed at removing or destroying the antigens. Antibodies have remarkable specificity, and the substitution of even one amino acid in a protein might prevent that protein from being recognized by an antibody.

For many bacterial species there are thousands of different strains (called serovars, for serological variants), which differ from one another mainly or solely in the antigenic identity of their lipopolysaccharide, flagella, or capsule. Different serovars of enteric bacteria—such as E. coli and Salmonella enterica, for example—are often found to be associated with the ability to inhabit different host animals or to cause different diseases. Formation of these numerous serovars reflects the ability of bacteria to respond effectively to the intense defensive actions of the immune system.

Bacterial Reproduction

Binary Fission

Most prokaryotes reproduce by a process of binary fission, in which the cell grows in volume until it divides in half to yield two identical daughter cells. Each daughter cell can continue to grow at the same rate as its parent. For this process to occur, the cell must grow over its entire surface until the time of cell division, when a new hemispherical pole forms at the division septum in the middle of the cell. In gram-positive bacteria the septum grows inward from the plasma membrane along the midpoint of the cell; in gram-negative bacteria the walls are more flexible, and the division septum forms as the side walls pinch inward, dividing the cell in two. In order for the cell to divide in half, the peptidoglycan structure must be different in the hemispherical cap than in the straight portion of the cell wall, and different wall-cross-linking enzymes must be active at the septum than elsewhere.

Budding

A group of environmental bacteria reproduces by budding. In this process a small bud forms at one end of the mother cell or on filaments called prosthecae. As growth proceeds, the size of the mother cell remains about constant, but the bud enlarges. When the bud is about the same size as the mother cell, it separates. This type of reproduction is analogous to that in budding fungi, such as brewer's yeast (Saccharomyces cerevisiae). One difference between fission and budding is that, in the latter, the mother cell often has different properties from the offspring. In some Pasteuria strains, the daughter buds have a flagellum and are motile, whereas the mother cells lack flagella but have long pili and holdfast appendages at the end opposite the bud. The related Planctomyces, found in plankton, have long fibrillar stalks at the end opposite the bud. In Hyphomicrobium a hyphal filament (prostheca) grows out of one end of the cell, and the bud grows out of the tip of the prostheca, separated by a relatively long distance from the mother cell.

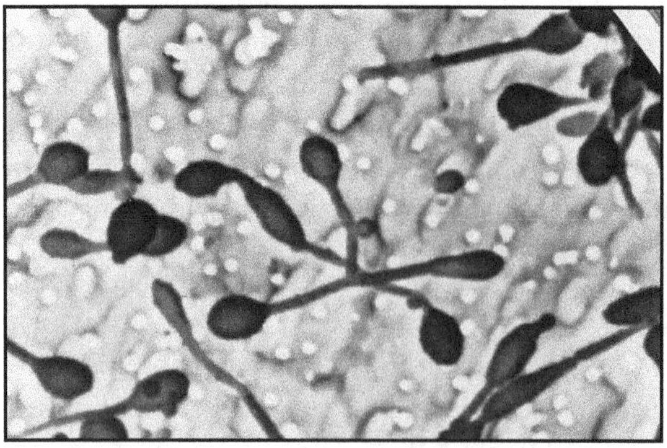

In the budding bacteria Hyphomonas polymorpha, the bud grows out of the end of a filament called a prostheca.

Sporulation

Many environmental bacteria are able to produce stable dormant, or resting, forms as a branch of their life cycle to enhance their survival under adverse conditions. These processes are not an obligate stage of the cell's life cycle but rather an interruption. Such dormant forms are called endospores, cysts, or heterocysts (primarily seen in cyanobacteria), depending on the method of spore formation, which differs between groups of bacteria.

The ability to form endospores is found among bacteria in a number of genera, predominantly gram-positive groups, including the aerobic rod Bacillus, the microaerophilic rod Sporolactobacillus, the anaerobic rods Clostridium and Desulfotomaculum, the coccus Sporosarcina, and the filamentous Thermoactinomyces. The formation of a spore occurs in response to nutritional deprivation. Consequently, endospores do not possess metabolic activity until nutrients become available, at which time they are able to differentiate from spores into vegetative cells. Only one spore is formed inside each bacterial cell during sporulation. The formation of a spore begins with invagination of the cytoplasmic membrane around a copy of the bacterial chromosome, thus separating the contents of the smaller cell from the mother cell. The membrane of the mother cell engulfs the smaller cell within its cytoplasm, effectively providing two concentric unit membranes to protect the developing spore. A thin spore membrane and a thick cortex of a peptidoglycan are laid down between the two unit membranes. A rigid spore coat forms outside the cortex, enclosing the entire spore structure. The spore coat has keratin-like properties that are able to resist the lethal effects of heat, desiccation (dehydration), freezing, chemicals, and radiation. The ability of endospores to resist these noxious agents may ensue from the extremely low water content inside the spore. Methane-oxidizing bacteria in the genus Methylosinus also produce desiccation-resistant spores, called exospores.

Cysts are thick-walled structures produced by dormant members of Azotobacter, Bdellovibrio (bdellocysts), and Myxococcus (myxospores). They are resistant to desiccation and other harmful conditions but to a lesser degree than are endospores. In encystment by the nitrogen-fixing Azotobacter, cell division is followed by the formation of a thick, multilayered wall and coat that surround the resting cell. The filamentous actinomycetes produce reproductive spores of two categories: conidiospores, which are chains of multiple spores formed on aerial or substrate mycelia, or sporangiospores, which are formed in specialized sacs called sporangia.

Exchange of Genetic Information

Bacteria do not have an obligate sexual reproductive stage in their life cycle, but they can be very active in the exchange of genetic information. The genetic information carried in the DNA can be transferred from one cell to another; however, this is not a true exchange, because only one partner receives the new information. In addition, the amount of DNA that is transferred is usually only a small piece of the chromosome. There are several mechanisms by which this takes place. In transformation, bacteria take up free fragments of DNA that are floating in the medium. To take up the DNA efficiently, bacterial cells must be in a competent state, which is defined by the capability of bacteria to bind free fragments of DNA and is formed naturally only in a limited number of bacteria, such as Haemophilus, Neisseria, Streptococcus, and Bacillus. Many other bacteria, including E. coli, can be rendered competent artificially under laboratory conditions, such as by exposure to solutions of calcium chloride (CaCl2). Transformation is a major tool in recombinant DNA technology, because fragments of DNA from one organism can be taken up by a second organism, thus allowing the second organism to acquire new characteristics.

Transduction is the transfer of DNA from one bacterium to another by means of a bacteria-infecting virus called a bacteriophage. Transduction is an efficient means of transferring DNA between bacteria because DNA enclosed in the bacteriophage is protected from physical decay and from attack by enzymes in the environment and is injected directly into cells by the bacteriophage. However, widespread gene transfer by means of transduction is of limited significance because the packaging of bacterial DNA into a virus is inefficient and the bacteriophages are usually highly restricted in the range of bacterial species that they can infect. Thus, interspecies transfer of DNA by transduction is rare.

Conjugation is the transfer of DNA by direct cell-to-cell contact that is mediated by plasmids (nonchromosomal DNA molecules). Conjugative plasmids encode an extremely efficient mechanism that mediates their own transfer from a donor cell to a recipient cell. The process takes place in one direction since only the donor cells contain the conjugative plasmid. In gram-negative bacteria, donor cells produce a specific plasmid-coded pilus, called the sex pilus, which attaches the donor cell to the recipient cell. Once connected, the two cells are brought into direct contact, and a conjugal bridge forms through which the DNA is transferred from the donor to the recipient. Many conjugative plasmids can be transferred between, and reproduce in, a large number of different gram-negative bacterial species. Plasmids vary in size, from a few thousand to more than 100,000 base pairs; the latter are sometimes called megaplasmids.

The bacterial chromosome can also be transferred during conjugation, although this happens less frequently than plasmid transfer. Conjugation allows the inheritance of large portions of genes and may be responsible for the existence of bacteria with traits of several different species. Conjugation also has been observed in the gram-positive genus Enterococcus, but the mechanism of cell recognition and DNA transfer is different from that which occurs in gram-negative bacteria.

Growth of Bacterial Populations

Growth of bacterial cultures is defined as an increase in the number of bacteria in a population rather than in the size of individual cells. The growth of a bacterial population occurs in a geometric or exponential manner: with each division cycle (generation), one cell gives rise to 2 cells, then 4 cells, then 8 cells, then 16, then 32, and so forth. The time required for the formation of a generation, the generation time (G), can be calculated from the following formula:

$$G = \frac{t}{n} = \frac{t}{3.33 \log b / B}$$

In the formula, B is the number of bacteria present at the start of the observation, b is the number present after the time period t, and n is the number of generations. The relationship shows that the mean generation time is constant and that the rate at which the number of bacteria increases is proportional to the number of bacteria at any given time. This relationship is valid only during the period when the population is increasing in an exponential manner, called the log phase of growth. For this reason, graphs that show the growth of bacterial cultures are plotted as the logarithm of the number of cells.

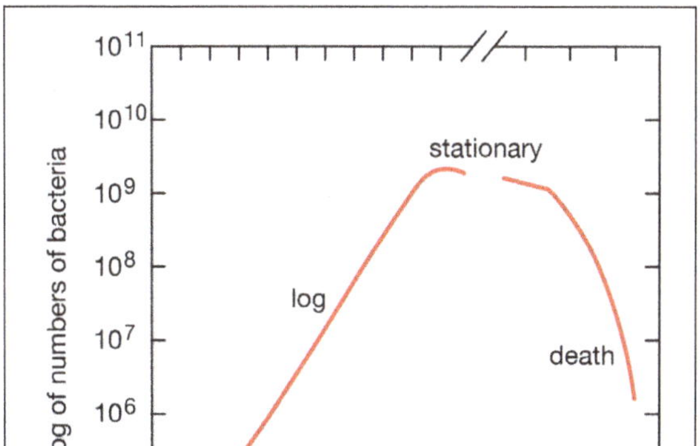

The generation time, which varies among bacteria, is controlled by many environmental conditions and by the nature of the bacterial species. For example, Clostridium perfringens, one of the fastest-growing bacteria, has an optimum generation time of about 10 minutes; Escherichia coli can double every 20 minutes; and the slow-growing Mycobacterium tuberculosis has a generation time in the range of 12 to 16 hours. Some researchers have suggested that certain bacteria populations living deep below Earth's surface may grow at extremely slow rates, reproducing just once every several thousand years. The composition of the growth medium is a major factor controlling the growth rate. The growth rate increases up to a maximum when the medium provides a better energy source and more of the biosynthetic intermediates that the cell would otherwise have to make for itself.

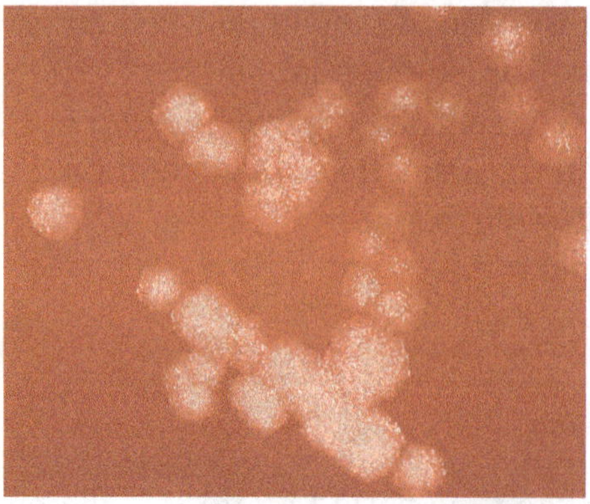

When bacteria are placed in a medium that provides all of the nutrients that are necessary for their growth, the population exhibits four phases of growth that are representative of a typical bacterial growth curve. Upon inoculation into the new medium, bacteria do not immediately reproduce, and the population size remains constant. During this period, called the lag phase, the cells are metabolically active and increase only in cell size. They are also synthesizing the enzymes and factors needed for cell division and population growth under their new environmental conditions. The population then enters the log phase, in which cell numbers increase in a logarithmic fashion, and each cell generation occurs in the same time interval as the preceding ones, resulting in a balanced increase in the constituents of each cell. The log phase continues until nutrients are depleted or toxic products accumulate, at which time the cell growth rate slows, and some cells may begin to die. Under optimum conditions, the maximum population for some bacterial species at the end of the log phase can reach a density of 10 to 30 billion cells per millilitre.

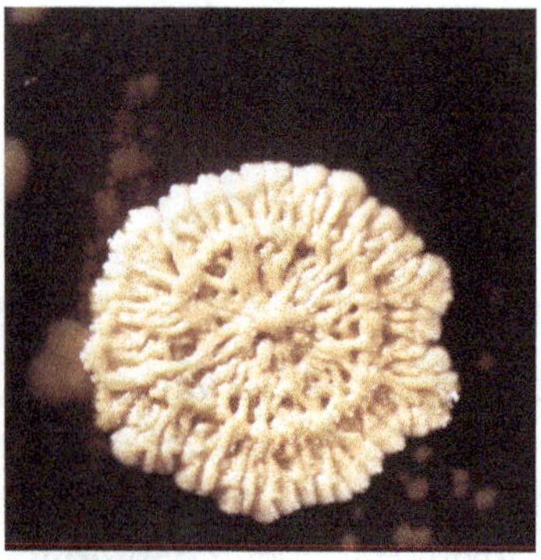

The log phase of bacterial growth is followed by the stationary phase, in which the size of a population of bacteria remains constant, even though some cells continue to divide and others begin to die. The stationary phase is followed by the death phase, in which the death of cells in the population exceeds the formation of new cells. The length of time before the onset of the death phase depends on the species and the medium. Bacteria do not necessarily die even when starved of nutrients, and they can remain viable for long periods of time.

MICROALGAE

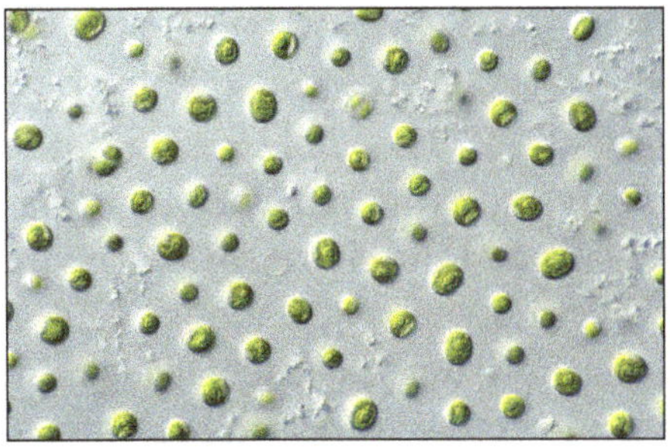

Microalgae CSIRO.

Collection of microalgae cultures, CSIRO.

Microalgae or microphytes are microscopic algae, typically found in freshwater and marine systems, living in both the water column and sediment. They are unicellular species which exist individually, or in chains or groups. Depending on the species, their sizes can range from a few micrometers (μm) to a few hundred micrometers. Unlike higher plants, microalgae do not have roots, stems, or leaves. They are specially adapted to an environment dominated by viscous forces. Microalgae, capable of performing photosynthesis, are important for life on earth; they produce approximately half of the atmospheric oxygen and use simultaneously the greenhouse gas carbon dioxide to grow photoautotrophically. Microalgae, together with bacteria, form the base of the food web and provide energy for all the trophic levels above them. Microalgae biomass is often measured with chlorophyll *a* concentrations and can provide a useful index of potential production. The standing stock of microphytes is closely related to that of its predators. Without grazing pressures the standing stock of microphytes dramatically decreases.

The biodiversity of microalgae is enormous and they represent an almost untapped resource. It has been estimated that about 200,000-800,000 species in many different genera exist of which about 50,000 species are described. Over 15,000 novel compounds originating from algal biomass have been chemically determined. Most of these microalgae species produce unique products like carotenoids, antioxidants, fatty acids, enzymes, polymers, peptides, toxins and sterols.

Characteristics and Uses

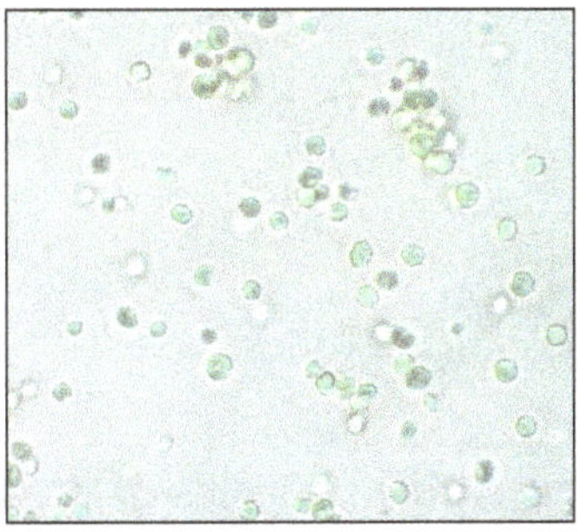

The microalgae *Nannochloropsis* sp., viewed under a light microscope.

The chemical composition of microalgae is not an intrinsic constant factor but varies over a wide range, both depending on species and on cultivation conditions. Some microalgae have the capacity to acclimate to changes in environmental conditions by altering their chemical composition in response to environmental variability. A particularly dramatic example is their ability to replace phospholipids with non-phosphorus membrane lipids in P-depleted environments. It is possible to accumulate the desired products in microalgae to a large extent by changing environmental factors, like temperature, illumination, pH, CO_2 supply, salt and nutrients. Microphytes also produce chemical signals which contribute to prey selection, defense, and avoidance. These chemical signals affect large scale tropic structures such as algal blooms but propagate by simple diffusion and laminar advective flow. Microalgae such as microphytes constitute the basic foodstuff for numerous aquaculture species, especially filtering bivalves. Photosynthetic and chemosynthetic microbes can also form symbiotic relationships with host organisms.

They provide them with vitamins and polyunsaturated fatty acids, necessary for the growth of the bivalves which are unable to synthesize it themselves.

In addition, because the cells grow in aqueous suspension, they have more efficient access to water, CO_2, and other nutrients. Microalgae play a major role in nutrient cycling and fixing inorganic carbon into organic molecules.

While fish oil has become famous for its omega-3 fatty acid content, fish don't actually produce omega-3s, instead accumulating their omega-3 reserves by consuming microalgae. These omega-3 fatty acids can be obtained in the human diet directly from the microalgae that produce them.

Aquaculture

A range of microalgae species are produced in hatcheries and are used in a variety of ways for commercial purposes.

Studies have estimated the main factors in the success of a microalgae hatchery system to be:

- The dimensions of the container/bioreactor where microalgae is cultured.
- Exposure to light/irradiation.
- Concentration of cells within the reactor.

FUNGUS

Fungus is any of about 144,000 known species of organisms of the kingdom Fungi, which includes the yeasts, rusts, smuts, mildews, molds, and mushrooms. There are also many funguslike organisms, including slime molds and oomycetes (water molds), that do not belong to kingdom Fungi but are often called fungi. Many of these funguslike organisms are included in the kingdom Chromista. Fungi are among the most widely distributed organisms on Earth and are of great environmental and medical importance. Many fungi are free-living in soil or water; others form parasitic or symbiotic relationships with plants or animals.

Fungi are eukaryotic organisms; i.e., their cells contain membrane-bound organelles and clearly defined nuclei. Historically, fungi were included in the plant kingdom; however, because fungi lack chlorophyll and are distinguished by unique structural and physiological features (i.e., components of the cell wall and cell membrane), they have been separated from plants. In addition, fungi are clearly distinguished from all other living organisms, including animals, by their principal modes of vegetative growth and nutrient intake. Fungi grow from the tips of filaments (hyphae) that make up the bodies of the organisms (mycelia), and they digest organic matter externally before absorbing it into their mycelia.

While mushrooms and toadstools (poisonous mushrooms) are by no means the most numerous or economically significant fungi, they are the most easily recognized. The Latin word for mushroom, fungus (plural fungi), has come to stand for the whole group. Similarly, the study of fungi is known as mycology—a broad application of the Greek word for mushroom, mykēs. Fungi other than mushrooms are sometimes collectively called molds, although this term is better restricted to fungi of the sort represented by bread mold.

Importance of Fungi

Humans have been indirectly aware of fungi since the first loaf of leavened bread was baked and the first tub of grape must was turned into wine. Ancient peoples were familiar with the ravages of fungi in agriculture but attributed these diseases to the wrath of the gods. The Romans designated a particular deity, Robigus, as the god of rust and, in an effort to appease him, organized an annual festival, the Robigalia, in his honour.

Fungi are everywhere in very large numbers—in the soil and the air, in lakes, rivers, and seas, on and within plants and animals, in food and clothing, and in the human body. Together with bacteria, fungi are responsible for breaking down organic matter and releasing carbon, oxygen, nitrogen, and phosphorus into the soil and the atmosphere. Fungi are essential to many household and industrial processes, notably the making of bread, wine, beer, and certain cheeses. Fungi are also used as food; for example, some mushrooms, morels, and truffles are epicurean delicacies, and mycoproteins (fungal proteins), derived from the mycelia of certain species of fungi, are used to make foods that are high in protein.

Studies of fungi have greatly contributed to the accumulation of fundamental knowledge in biology. For example, studies of ordinary baker's or brewer's yeast (Saccharomyces cerevisiae) led to discoveries of basic cellular biochemistry and metabolism. Some of these pioneering discoveries were made at the end of the 19th century and continued during the first half of the 20th century. From 1920 through the 1940s, geneticists and biochemists who studied mutants of the red bread mold, Neurospora, established the one-gene–one-enzyme theory, thus contributing to the foundation of modern genetics. Fungi continue to be useful for studying cell and molecular biology, genetic engineering, and other basic disciplines of biology.

The medical relevance of fungi was discovered in 1928, when Scottish bacteriologist Alexander Fleming noticed the green mold Penicillium notatum growing in a culture dish of Staphylococcus bacteria. Around the spot of mold was a clear ring in which no bacteria grew. Fleming successfully isolated the substance from the mold that inhibited the growth of bacteria. In 1929 he published a scientific report announcing the discovery of penicillin, the first of a series of antibiotics—many of them derived from fungi—that have revolutionized medical practice.

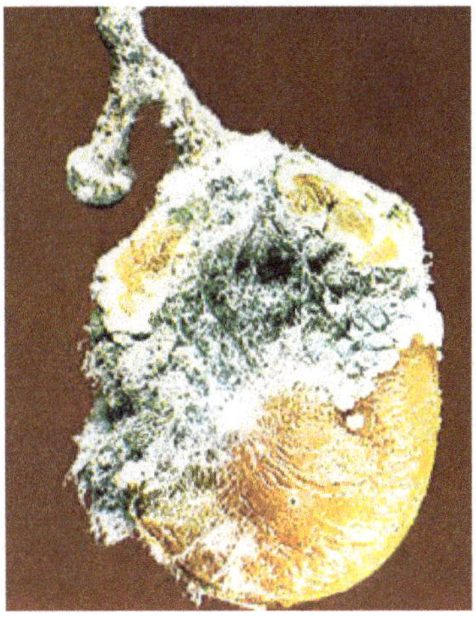

Another medically important fungus is Claviceps purpurea, which is commonly called ergot and causes a plant disease of the same name. The disease is characterized by a growth that develops on grasses, especially on rye. Ergot is a source of several chemicals used in drugs that induce labour in pregnant women and that control hemorrhage after birth. Ergot is also the source of lysergic acid, the active principle of the psychedelic drug lysergic acid diethylamide (LSD). Other species of fungi contain chemicals that are extracted and used to produce drugs known as statins, which control cholesterol levels and ward off coronary heart disease. Fungi are also used in the production of a number of organic acids, enzymes, and vitamins.

Form and Function of Fungi

Size Range

The mushrooms, because of their size, are easily seen in fields and forests and consequently were the only fungi known before the invention of the microscope in the 17th century. The microscope made it possible to recognize and identify the great variety of fungal species living on dead or live organic matter. The part of a fungus that is generally visible is the fruiting body, or sporophore. Sporophores vary greatly in size, shape, colour, and longevity. Some are microscopic and completely invisible to the unaided eye; others are no larger than a pin head; still others are gigantic structures. Among the largest sporophores are those of mushrooms, bracket fungi, and puffballs. Some mushrooms reach a diameter of 20 to 25 cm (8 to 10 inches) and a height of 25 to 30 cm (10 to 12 inches). Bracket, or shelf, fungi can reach 40 cm (16 inches) or more in diameter. A specimen of the bracket fungus Fomitiporia ellipsoidea discovered in 2010 on Hainan Island in southern China had a fruiting body measuring 10.8 metres (35.4 feet) in length and 82–88 cm (2.7–2.9 feet) in width. It may have held some 450 million spores and weighed an estimated 400–500 kg (882–1,102 pounds), at the time making it the largest fungal fruiting body ever documented. Puffballs also can grow to impressive sizes. The largest puffballs on record measured 150 cm (5 feet) in diameter. The number of spores within such giants reaches several trillion.

Bracket fungi, which grow on tree trunks, are among some of the largest fungi. Some species may reach 40 cm (16 inches) in diameter.*H.S. Knighton.*

Distribution and Abundance

Fungi are either terrestrial or aquatic, the latter living in freshwater or marine environments. Freshwater species are usually found in clean, cool water because they do not tolerate high degrees of salinity. However, some species are found in slightly brackish water, and a few thrive in highly polluted streams. Soil that is rich in organic matter furnishes an ideal habitat for a large number of species; only a small number of species are found in drier areas or in habitats with little or no organic matter. Fungi are found in all temperate and tropical regions of the world where there is sufficient moisture to enable them to grow. A few species of fungi live in the Arctic and Antarctic regions, although they are rare and are more often found living in symbiosis with algae in the form of lichens About 144,000 species of fungi have been identified and described, but mycologists estimate that there may be between 2.2 million and 3.8 million total species.

Basic Morphology

A typical fungus consists of a mass of branched, tubular filaments enclosed by a rigid cell wall. The filaments, called hyphae (singular hypha), branch repeatedly into a complicated, radially expanding network called the mycelium, which makes up the thallus, or undifferentiated body, of the typical fungus. The mycelium grows by utilizing nutrients from the environment and, upon reaching a certain stage of maturity, forms—either directly or in special fruiting bodies—reproductive cells called spores. The spores are released and dispersed by a wide variety of passive or active mechanisms; upon reaching a suitable substrate, the spores germinate and develop hyphae that grow, branch repeatedly, and become the mycelium of the new individual. Fungal growth is mainly confined to the tips of the hyphae, and all fungal structures are therefore made up of hyphae or portions of hyphae.

Shelf fungi Fungi are made up of masses of tubular filaments called hyphae that penetrate into and absorb nutrients from the substrates on which fungi grow. Some fungi have extensive networks of hyphae that enable the fruiting body of the fungi to grow very large, such as many species of shelf, or bracket, fungi.

Some fungi, notably the yeasts, do not form a mycelium but grow as individual cells that multiply by budding or, in certain species, by fission. In addition, the so-called cryptomycota, a primitive group of microscopic fungi, diverge significantly from the standard body plan of other fungi in that their cell walls lack the rigid polymer known as chitin. These microscopic fungi also possess a whiplike flagellum.

Structure of the Thallus

In almost all fungi the hyphae that make up the thallus have cell walls. (The thalli of the true slime molds lack cell walls and, for this and other reasons, are classified as protists rather than fungi.) A hypha is a multibranched tubular cell filled with cytoplasm. The tube itself may be either continuous throughout or divided into compartments, or cells, by cross walls called septa (singular septum). In nonseptate (i.e., coenocytic) hyphae the nuclei are scattered throughout the cytoplasm. In septate hyphae each cell may contain one to many nuclei, depending on the type of fungus or the stage of hyphal development. The cells of fungi are similar in structure to those of many other organisms. The minute nucleus, readily seen only in young portions of the hypha, is surrounded by a double membrane and typically contains one nucleolus. In addition to the nucleus, various organelles—such as the endoplasmic reticulum, Golgi apparatus, ribosomes, and liposomes—are scattered throughout the cytoplasm.

The thalli of fungi, which are hidden underground in soil fungi such as Amanita, are made up of mycelia and lack specialized tissues.

USDA Forest Service

Hyphae usually are either nonseptate (generally in the more primitive fungi) or incompletely septate (meaning that the septa are perforated). This permits the movement of cytoplasm (cytoplasmic streaming) from one cell to the next. In fungi with perforated septa, various molecules are able to move rapidly between hyphal cells, but the movement of larger organelles, such as mitochondria and nuclei, is prevented. In the absence of septa, both mitochondria and nuclei can be readily translocated along hyphae. In mating interactions between filamentous Basidiomycota, the nuclei of one parent often invade the hyphae of the other parent, because the septa are degraded ahead of the incoming nuclei to allow their passage through the existing hyphae. Once the incoming nuclei are established, septa are re-formed.

Variations in the structure of septa are numerous in the fungi. Some fungi have sievelike septa called pseudosepta, whereas fungi in other groups have septa with one to few pores that are small enough in size to prevent the movement of nuclei to adjacent cells. Basidiomycota have a septal structure called a dolipore septum that is composed of a pore cap surrounding a septal swelling and septal pore. This organization permits cytoplasm and small organelles to pass through but restricts the movement of nuclei to varying degrees.

Club fungus (Claveria) has a dolipore septum, which is made up of a pore cap that surrounds a septal swelling and septal pore.Ken Brate/Photo Researchers.

The wall of the hypha is complex in both composition and structure. Its exact chemical composition varies in different fungal groups. In some funguslike organisms the wall contains considerable quantities of cellulose, a complex carbohydrate that is the chief constituent of the cell walls of plants. In most fungi, however, two other polymers—chitin and glucan (a polymer of glucose linked at the third carbon and branched at the sixth), which forms an α-glucan layer and a special β-1,3-1,6-glucan layer—form the main structural components of the wall. Among the many other chemical substances in the walls of fungi are some that may thicken or toughen the wall of tissues, thus imparting rigidity and strength. The chemical composition of the wall of a particular fungus may vary at different stages of the organism's growth—a possible indication that the wall plays some part in determining the form of the fungus. In some fungi, carbohydrates are stored in the wall at one stage of development and are removed and utilized at a later stage. In some yeasts, fusion of sexually functioning cells is brought about by the interaction of specific chemical substances on the walls of two compatible mating types.

When the mycelium grows in or on a surface, such as in the soil, on a log, or in culture medium, it appears as a mass of loose, cottony threads. The richer the composition of the growth medium, the more profuse the threads and the more feltlike the mass. On the sugar-rich growth substances used in laboratories, the assimilative (somatic) hyphae are so interwoven as to form a thick, almost leathery colony. On the soil, inside a leaf, in the skin of animals, or in other parasitized plant or animal tissues, the hyphae are usually spread in a loose network. The mycelia of the so-called higher fungi does, however, become organized at times into compact masses of different sizes that serve various functions. Some of these masses, called sclerotia, become extremely hard and serve to carry the fungus over periods of adverse conditions of temperature and moisture. One example of a fungus that forms sclerotia is ergot (Claviceps purpurea), which causes a disease of cereal grasses. The underground sclerotia of Wolfiporia extensa, an edible pore fungus also known as tuckahoe, may reach a diameter of 20 to 25 cm (8 to 10 inches).

Various other tissues are also produced by the interweaving of the assimilative hyphae of some fungi. Stromata (singular stroma) are cushionlike tissues that bear spores in various ways. Rhizomorphs are long strands of parallel hyphae cemented together. Those of the honey mushroom (Armillaria mellea), which are black and resemble shoestrings, are intricately constructed and are differentiated to conduct water and food materials from one part of the thallus to another.

Sporophores and Spores

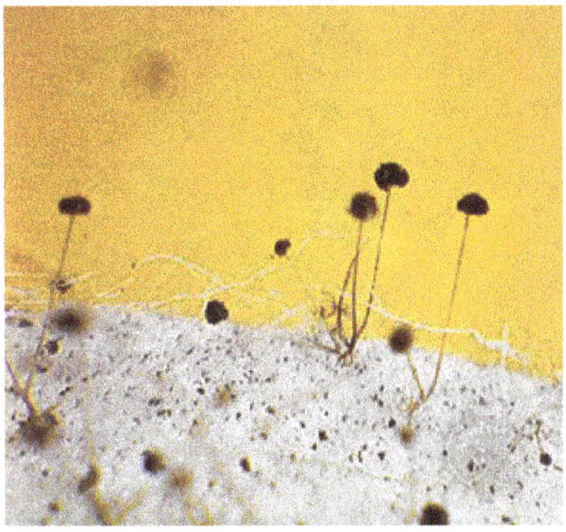

When the mycelium of a fungus reaches a certain stage of growth, it begins to produce spores either directly on the somatic hyphae or, more often, on special sporiferous (spore-producing) hyphae, which may be loosely arranged or grouped into intricate structures called fruiting bodies, or sporophores.

Fungi can produce spores on somatic hyphae or on special spore-producing hyphae, which are arranged into structures called fruiting bodies. Earth tongue fungi (Geoglossum fallax) have club-shaped fruiting bodies and produce ascospores in sacs called asci.

The more primitive fungi produce spores in sporangia, which are saclike sporophores whose entire cytoplasmic contents cleave into spores, called sporangiospores. Thus, they differ from more advanced fungi in that their asexual spores are endogenous. Sporangiospores are either naked and flagellated (zoospores) or walled and nonmotile (aplanospores). The more primitive aquatic and terrestrial fungi tend to produce zoospores. The zoospores of aquatic fungi and funguslike organisms swim in the surrounding water by means of one or two variously located flagella (whiplike organs of locomotion). Zoospores produced by terrestrial fungi are released after a rain from the sporangia in which they are borne and swim for a time in the rainwater between soil particles or on the wet surfaces of plants, where the sporangia are formed by parasitic fungi. After some time, the zoospores lose their flagella, surround themselves with walls, and encyst. Each cyst germinates by producing a germ tube. The germ tube may develop a mycelium or a reproductive structure, depending on the species and on the environmental conditions. The bread molds, which are the most advanced of the primitive fungi, produce only aplanospores (nonmotile spores) in their sporangia.

Crucibulum vulgare, a species of bird's-nest fungi, forms egg-shaped basidiocarps and is a member of the phylum Basidiomycota.

The more advanced fungi do not produce motile spores of any kind, even though some of them are aquatic in fresh or marine waters. In these fungi, asexually produced spores (usually called conidia) are produced exogenously and are typically formed terminally or laterally on special spore-producing hyphae called conidiophores. Conidiophores may be arranged singly on the hyphae or may be grouped in special asexual fruiting bodies, such as flask-shaped pycnidia, mattresslike acervuli, cushion-shaped sporodochia, or Sexually produced spores of the higher fungi result from meiosis and are formed either in saclike structures (asci) typical of the Ascomycota or on the surface of club-shaped structures (basidia) typical of the Basidiomycota. Asci and basidia may be borne naked, directly on the hyphae, or in various types of sporophores, called ascocarps (also known as ascomata) or basidiocarps (also known as basidiomata), depending on whether they bear asci or basidia, respectively. Well-known examples of ascocarps are the morels, the cup fungi, and the truffles. Commonly encountered basidiocarps are mushrooms, brackets, puffballs, stinkhorns, and bird's-nest fungi.

Growth

Under favourable environmental conditions, fungal spores germinate and form hyphae. During this process, the spore absorbs water through its wall, the cytoplasm becomes activated, nuclear division takes place, and more cytoplasm is synthesized. The wall initially grows as a spherical structure. Once polarity is established, a hyphal apex forms, and from the wall of the spore a germ tube bulges out, enveloped by a wall of its own that is formed as the germ tube grows.

The hypha may be roughly divided into three regions: (1) the apical zone about 5–10 micrometres (0.0002–0.0004 inch) in length, (2) the subapical region, extending about 40 micrometres (0.002 inch) back of the apical zone, which is rich in cytoplasmic components, such as nuclei, Golgi apparatus, ribosomes, mitochondria, the endoplasmic reticulum, and vesicles, but is devoid of vacuoles, and (3) the zone of vacuolation, which is characterized by the presence of many vacuoles and the accumulation of lipids.

Growth of hyphae in most fungi takes place almost exclusively in the apical zone (i.e., at the very tip). This is the region where the cell wall extends continuously to produce a long hyphal tube. The cytoplasm within the apical zone is filled with numerous vesicles. These bubblelike structures are usually too small to be seen with an ordinary microscope but are clearly evident under the electron microscope. In higher fungi the apical vesicles can be detected with an ordinary microscope equipped with phase-contrast optics as a round spot with a somewhat diffuse boundary. This body is universally known by its German name, the Spitzenkörper, and its position determines the direction of growth of a hypha.

The growing tip eventually gives rise to a branch. This is the beginning of the branched mycelium. Growing tips that come in contact with neighbouring hyphae often fuse with them to form a hyphal net. In such a vigorously growing system, the cytoplasm is in constant motion, streaming toward the growing tips. Eventually, the older hyphae become highly vacuolated and may be stripped of most of their cytoplasm. All living portions of a thallus are potentially capable of growth. If a small piece of mycelium is placed under conditions favourable for growth, it develops into a new thallus, even if no growing tips are included in the severed portion.

Growth of a septate mycelium (i.e., with cross walls between adjacent cells) entails the formation of new septa in the young hyphae. Septa are formed by ringlike growth from the wall of the hypha toward the centre until the septa are complete. In the higher fungi the septum stops growing before it is complete; the result is a central pore through which the cytoplasm flows, thus establishing organic connection throughout the thallus. In contrast to plants, in which the position of the septum separating two daughter cells determines the formation of tissues, the fungal septum is always formed at right angles to the axis of growth. As a result, in fungal tissue formation, the creation of parallel hyphae cannot result from longitudinal septum formation but only from outgrowth of a new branch. In fungi, therefore, the mechanism that determines the point of origin and subsequent direction of growth of hyphal branches is the determining factor in developmental morphogenesis.

The individual fungus is potentially immortal, because it continues to grow at the hyphal tips as long as conditions remain favourable. It is possible that, in undisturbed places, mycelia exist that have grown continuously for many thousands of years. The older parts of the hyphae die and decompose, releasing nitrogen and other nutrients into the soil.

Some species of endophytic fungi, such as Neotyphodium and Epichloë, which invade the seeds of grasses (e.g., ryegrass and fescue) and grow within the plant, grow not through extension of the hyphal tips but by intercalary growth, in which the hyphae attach to the growing cells of the plant. This type of growth enables the hyphae of the fungus to grow at the same rate that the plant grows. Intercalary growth of endophytic fungi was discovered in 2007, although for many years scientists suspected that these fungi possessed unique adaptations that allow them to grow as if they were natural parts of their hosts.

The underground network of hyphae of a mushroom can grow and spread over a very large area, often several metres (yards) in diameter. The underground hyphae obtain food from organic matter in the substratum and grow outward. The hyphal branches at the edge of the mycelium become organized at intervals into elaborate tissues that develop aboveground into mushrooms. Such a circle of mushrooms is known as a fairy ring, because in the Middle Ages it was believed to represent the path of dancing fairies. The ring marks the periphery of an enormous fungus colony, which, if undisturbed, continues to produce ever wider fairy rings year after year. Fungi can grow into enormous colonies. Some thalli of Armillaria species, which are pathogens of forest trees, are among the largest and oldest organisms on Earth.

Amanita alba, a species of mushroom, can grow into a large fungal colony that forms a circle of mushrooms known as a fairy ring.

Nutrition

Unlike plants, which use carbon dioxide and light as sources of carbon and energy, respectively, fungi meet these two requirements by assimilating preformed organic matter; carbohydrates are generally the preferred carbon source. Fungi can readily absorb and metabolize a variety of soluble carbohydrates, such as glucose, xylose, sucrose, and fructose. Fungi are also characteristically well equipped to use insoluble carbohydrates such as starches, cellulose, and hemicelluloses, as well as very complex hydrocarbons such as lignin. Many fungi can also use proteins as a source of carbon and nitrogen. To use insoluble carbohydrates and proteins, fungi must first digest these polymers extracellularly. Saprotrophic fungi obtain their food from dead organic material; parasitic fungi do so by feeding on living organisms (usually plants), thus causing disease.

Fungi secure food through the action of enzymes (biological catalysts) secreted into the surface on which they are growing; the enzymes digest the food, which then is absorbed directly through the hyphal walls. Food must be in solution in order to enter the hyphae, and the entire mycelial surface of a fungus is capable of absorbing materials dissolved in water. The rotting of fruits, such as peaches and citrus fruits in storage, demonstrates this phenomenon, in which the infected parts are softened by the action of the fungal enzymes. In brown rot of peaches, the softened area is somewhat larger than the actual area invaded by the hyphae: the periphery of the brown spot has been softened by enzymes that act ahead of the invading mycelium. Cheeses such as Brie and Camembert are matured by enzymes produced by the fungus Penicillium camemberti, which grows on the outer surface of some cheeses. Some fungi produce special rootlike hyphae, called rhizoids, which anchor the thallus to the growth surface and probably also absorb food. Many parasitic fungi are even more specialized in this respect, producing special absorptive organs called haustoria.

Saprotrophism

Together with bacteria, saprotrophic fungi are to a large extent responsible for the decomposition of organic matter. They are also responsible for the decay and decomposition of foodstuffs. Among other destructive saprotrophs are fungi that destroy timber and timber products as their mycelia invade and digest the wood; many of these fungi produce their spores in large, woody, fruiting bodies—e.g., bracket or shelf fungi. Paper, textiles, and leather are often attacked and destroyed by fungi. This is particularly true in tropical regions, where temperature and humidity are often very high.

Many fungi are saprobic—i.e., they obtain nutrients from dead organic matter.

The nutritional requirements of saprotrophs (and of some parasites that can be cultivated artificially) have been determined by growing fungi experimentally on various synthetic substances of known chemical composition. Fungi usually exhibit the same morphological characteristics in these culture media as they do in nature. Carbon is supplied in the form of sugars or starch; the majority of fungi thrive on such sugars as glucose, fructose, mannose, maltose, and, to a lesser extent, sucrose. Decomposition products of proteins, such as proteoses, peptones, and amino acids, can be used by most fungi as nitrogen sources; ammonium compounds and nitrates also serve as nutrients for many species. It is doubtful, however, that any fungus can combine, or fix, atmospheric nitrogen into usable compounds. Chemical elements such as phosphorus, sulfur, potassium, magnesium, and small quantities of iron, zinc, manganese, and copper are needed by most fungi for vigorous growth; elements such as calcium, molybdenum, and gallium are required by at least some species. Oxygen and hydrogen are absolute requirements; they are supplied in the form of water or are obtained from carbohydrates. Many fungi, deficient in thiamine and biotin, must obtain these vitamins from the environment; most fungi appear able to synthesize all other vitamins necessary for their growth and reproduction.

As a rule, fungi are aerobic organisms, meaning they require free oxygen in order to live. Fermentations, however, take place under anaerobic conditions. Knowledge of the physiology of saprotrophic fungi has enabled industry to use several species for fermentation purposes. One of the most important groups of strictly anaerobic fungi are members of the genera Neocallimastix (phylum Neocallimastigomycota), which form a crucial component of the microbial population of the rumen of herbivorous mammals. These fungi are able to degrade plant cell wall components, such as cellulose and xylans, that the animals cannot otherwise digest.

Parasitism in Plants and Insects

In contrast with the saprotrophic fungi, parasitic fungi attack living organisms, penetrate their outer defenses, invade them, and obtain nourishment from living cytoplasm, thereby causing disease and sometimes death of the host. Most pathogenic (disease-causing) fungi are parasites of plants. Most parasites enter the host through a natural opening, such as a stoma (microscopic air pore) in a leaf, a lenticel (small opening through bark) in a stem, a broken plant hair or a hair socket in a fruit, or a wound in the plant. Among the most common and widespread diseases of plants caused by fungi are the various downy mildews (e.g., of grape, onion, tobacco), the powdery mildews (e.g., of grape, cherry, apple, peach, rose, lilac), the smuts (e.g., of corn, wheat, onion), the rusts (e.g., of wheat, oats, beans, asparagus, snapdragon, hollyhock), apple scab, brown rot of stone fruits, and various leaf spots, blights, and wilts. These diseases cause great damage annually throughout the world, destroying many crops and other sources of food. For example, nearly all the chestnut forests of the United States have been destroyed by the chestnut blight fungus (Cryphonectria parasitica), and the elms in both the United States and Europe have been devastated by Ophiostoma ulmi, the fungus that causes Dutch elm disease.

Infection of a plant takes place when the spores of a pathogenic fungus fall on the leaves or the stem of a susceptible host and germinate, each spore producing a germ tube. The tube grows on the surface of the host until it finds an opening; then the tube enters the host, puts out branches between the cells of the host, and forms a mycelial network within the invaded tissue. The germ tubes of some fungi produce special pressing organs called appressoria, from which a microscopic, needlelike peg presses against and punctures the epidermis of the host; after penetration, a mycelium develops in the usual manner. Many parasitic fungi absorb food from the host cells through the hyphal walls appressed against the cell walls of the host's internal tissues. Others produce haustoria (special absorbing structures) that branch off from the intercellular hyphae and penetrate the cells themselves. Haustoria, which may be short, bulbous protrusions or large branched systems filling the whole cell, are characteristically produced by obligate (i.e., invariably parasitic) parasites; some facultative (i.e., occasionally parasitic) parasites also produce them. Obligate parasites, which require living cytoplasm and have extremely specialized nutritional requirements, are exceptionally difficult, and often impossible, to grow in a culture dish in a laboratory. Examples of obligate parasites are the downy mildews, the powdery mildews, and the rusts.

Certain fungi form highly specialized parasitic relationships with insects. For example, the fungal genus Septobasidium is parasitic on scale insects (order Homoptera) that feed on trees. The mycelium forms elaborate structures over colonies of insects feeding on the bark. Each insect sinks its proboscis (tubular sucking organ) into the bark and remains there the rest of its life, sucking sap. The fungus sinks haustoria into the bodies of some of the insects and feeds on them without killing them. The parasitized insects are, however, rendered sterile. The perpetuation of the insect species and the spread of the fungus are accomplished by the uninfected members of the colony, which are protected from enemies by the fungus body. Newly hatched scale insects crawl over the surface of the fungus, which is at that time sporulating. Fungal spores adhere to the young insects and germinate. As the young insects settle down in a new place on the bark to begin feeding, they establish new fungal colonies. Thus, part of the insect colony is sacrificed to the fungus as food in return for the fungal protection provided for the rest of the insects. The insect is parasitic on the tree and the fungus is parasitic on the insect, but the tree is the ultimate victim.

Some fungi are parasitic on insects. For example, Cordyceps militaris invades living insect pupa by drawing nutrients from the pupa that enable the fungus to grow and generate spores for reproduction.

The sooty molds constitute another interesting ecological group of fungi that are associated with insects. The majority of sooty molds are tropical or subtropical, but some species occur in the temperate zones. All sooty molds are epiphytic (i.e., they grow on the surfaces of other plants), but only in areas where scale insects are present. The fungi parasitize neither the plants nor the insects but rather obtain their nourishment exclusively from the honeydew secretions of the scale insects. Growth of the dark mycelium over the plant leaves, however, is often so dense as to significantly reduce the intensity of the light that reaches the leaf surface; this reduction in turn significantly reduces the rate of photosynthesis. Insect-fungus associations found in the tropical forests of Central and South America include the unique relationship of leafcutter ants (sometimes called parasol ants) with fungi in the family Lepiotaceae (phylum Basidiomycota). The ants cultivate the fungi in their nests as an ongoing food supply and secrete enzymes that stimulate or suppress the growth of the fungi.

Parasitism in Humans

Many pathogenic fungi are parasitic in humans and are known to cause diseases of humans and other animals. In humans, parasitic fungi most commonly enter the body through a wound in the epidermis (skin). Such wounds may be insect punctures or accidentally inflicted scratches, cuts, or bruises. One example of a fungus that causes disease in humans is Claviceps purpurea, the cause of ergotism (also known as St. Anthony's fire), a disease that was prevalent in northern Europe in the Middle Ages, particularly in regions of high rye-bread consumption. The wind carries the fungal spores of ergot to the flowers of the rye, where the spores germinate, infect and destroy the ovaries of the plant, and replace them with masses of microscopic threads cemented together into a hard fungal structure shaped like a rye kernel but considerably larger and darker. This structure, called an ergot, contains a number of poisonous organic compounds called alkaloids. A mature head of rye may carry several ergots in addition to noninfected kernels. When the grain is harvested, much of the ergot falls to the ground, but some remains on the plants and is mixed with the grain. Although modern grain-cleaning and milling methods have practically eliminated the disease, the contaminated flour may end up in bread and other food products if the ergot is not removed before milling. In addition, the ergot that falls to the ground may be consumed by cattle turned out to graze in rye fields after harvest. Cattle that consume enough ergot may suffer abortion of fetuses or death. In the spring, when the rye is in bloom, the ergot remaining on the ground produces tiny, black, mushroom-shaped bodies that expel large numbers of spores, thus starting a new series of infections.

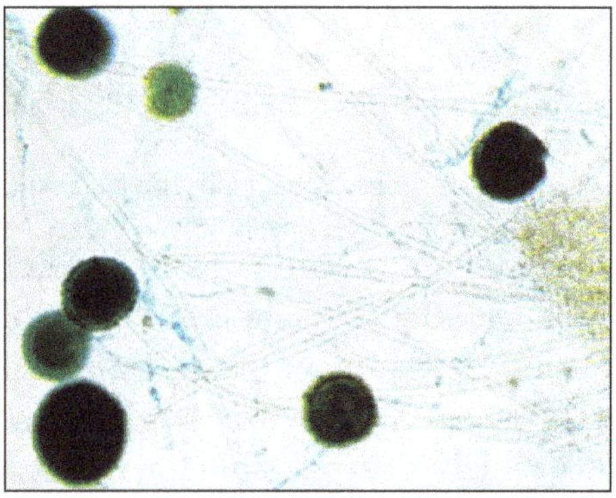

Some species of the fungi genus Aspergillus can cause allergic reactions and mild pneumonia in susceptible individuals.

Other human diseases caused by fungi include athlete's foot, ringworm, aspergillosis, histoplasmosis, and coccidioidomycosis. The yeast Candida albicans, a normal inhabitant of the human mouth, throat, colon, and reproductive organs, does not cause disease when it is in ecological balance with other microbes of the digestive system. However, disease, age, and hormonal changes can cause C. albicans to grow in a manner that cannot be controlled by the body's defense systems, resulting in candidiasis (called thrush when affecting the mouth). Candidiasis is characterized by symptoms ranging from irritating inflamed patches on the skin or raised white patches on the tongue to life-threatening invasive infection that damages the lining of the heart or brain. Improved diagnosis and increased international travel, the latter of which has facilitated the spread of tropical pathogenic fungi, have resulted in an increased incidence of fungal disease in humans. In addition, drug therapies used to manage the immune system in transplant and cancer patients weaken the body's defenses against fungal pathogens. Patients infected with human immunodeficiency virus (HIV), the causative agent of acquired immunodeficiency syndrome (AIDS), have similarly weakened immune defenses against fungi, and many AIDS-related deaths are caused by fungal infections (especially infection with Aspergillus fumigatus).

Mycorrhiza

Pinesaps (Monotropa hypopitys) are saprophytes with little photosynthetic tissue. Unlike green plants, saprophytes are unable to manufacture carbohydrates. They rely on their associations with mycorrhizal fungi, which synthesize carbohydrates from the rich organic leaf litter.

Among symbiotic fungi, those that enter into mycorrhizal relationships and those that enter into relationships with algae to form lichens are probably the best-known. A large number of fungi infect the roots of plants by forming an association with plants called mycorrhiza (plural mycorrhizas or mycorrhizae). This association differs markedly from ordinary root infection, which is responsible for root rot diseases. Mycorrhiza is a non-disease-producing association in which the fungus invades the root to absorb nutrients. Mycorrhizal fungi establish a mild form of parasitism that is mutualistic, meaning both the plant and the fungus benefit from the association. About 90 percent of land plants rely on mycorrhizal fungi, especially for mineral nutrients (i.e., phosphorus), and in return the fungus receives nutrients formed by the plant. During winter, when day length is shortened and exposure to sunlight is reduced, some plants produce few or no nutrients and thus depend on fungi for sugars, nitrogenous compounds, and other nutrients that the fungi are able to absorb from waste materials in the soil. By sharing the products it absorbs from the soil with its plant host, a fungus can keep its host alive. In some lowland forests, the soil contains an abundance of mycorrhizal fungi, resulting in mycelial networks that connect the trees together. The trees and their seedlings can use the fungal mycelium to exchange nutrients and chemical messages.

There are two main types of mycorrhiza: ectomycorrhizae and endomycorrhizae. Ectomycorrhizae are fungi that are only externally associated with the plant root, whereas endomycorrhizae form their associations within the cells of the host.

Among the mycorrhizal fungi are boletes, whose mycorrhizal relationships with larch trees (Larix) and other conifers have long been known. Other examples include truffles, some of which are believed to form mycorrhizae with oak (Quercus) or beech (Fagus) trees. Many orchids form mycorrhizae with species of Rhizoctonia that provide seedlings of the orchid host with carbohydrate obtained by degradation of organic matter in the soil.

Predation

A number of fungi have developed ingenious mechanisms for trapping microorganisms such as amoebas, roundworms (nematodes), and rotifers. After the prey is captured, the fungus uses hyphae to penetrate and quickly destroy the prey. Many of these fungi secrete adhesive substances over the surface of their hyphae, causing a passing animal that touches any portion of the mycelium to adhere firmly to the hyphae. For example, the mycelia of oyster mushrooms (genus Pleurotus) secrete adhesives onto their hyphae in order to catch nematodes. Once a passing animal is caught, a penetration tube grows out of a hypha and penetrates the host's soft body. This haustorium grows and branches and then secretes enzymes that quickly kill the animal, whose cytoplasm serves as food for the fungus.

Other fungi produce hyphal loops that ensnare small animals, thereby allowing the fungus to use its haustoria to penetrate and kill a trapped animal. Perhaps the most amazing of these fungal traps are the so-called constricting rings of some species of Arthrobotrys, Dactylella, and Dactylaria—soil-inhabiting fungi easily grown under laboratory conditions. In the presence of nematodes, the mycelium produces large numbers of rings through which the average nematode is barely able to pass. When a nematode rubs the inner wall of a ring, which usually consists of three cells with touch-sensitive inner surfaces, the cells of the ring swell rapidly, and the resulting constriction holds the worm tightly. All efforts of the nematode to free itself fail, and a hypha, which grows out of one of the swollen ring cells at its point of contact with the worm, penetrates and branches within the animal's body, thereby killing the animal. The dead animal is then used for food by the fungus. In the absence of nematodes, these fungi do not usually produce rings in appreciable quantities. A substance secreted by nematodes stimulates the fungus to form the mycelial rings.

Reproductive Processes of Fungi

Following a period of intensive growth, fungi enter a reproductive phase by forming and releasing vast quantities of spores. Spores are usually single cells produced by fragmentation of the mycelium or within specialized structures (sporangia, gametangia, sporophores, etc.). Spores may be produced either directly by asexual methods or indirectly by sexual reproduction. Sexual reproduction in fungi, as in other living organisms, involves the fusion of two nuclei that are brought together when two sex cells (gametes) unite. Asexual reproduction, which is simpler and more direct, may be accomplished by various methods.

Asexual Reproduction

Typically in asexual reproduction, a single individual gives rise to a genetic duplicate of the progenitor without a genetic contribution from another individual. Perhaps the simplest method of reproduction of fungi is by fragmentation of the thallus, the body of a fungus. Some yeasts, which are single-celled fungi, reproduce by simple cell division, or fission, in which one cell undergoes nuclear division and splits into two daughter cells; after some growth, these cells divide, and eventually a population of cells forms. In filamentous fungi the mycelium may fragment into a number of segments, each of which is capable of growing into a new individual. In the laboratory, fungi are commonly propagated on a layer of solid nutrient agar inoculated either with spores or with fragments of mycelium.

Budding, which is another method of asexual reproduction, occurs in most yeasts and in some filamentous fungi. In this process, a bud develops on the surface of either the yeast cell or the hypha, with the cytoplasm of the bud being continuous with that of the parent cell. The nucleus of the parent cell then divides; one of the daughter nuclei migrates into the bud, and the other remains in the parent cell. The parent cell is capable of producing many buds over its surface by continuous synthesis of cytoplasm and repeated nuclear divisions. After a bud develops to a certain point and even before it is severed from the parent cell, it is itself capable of budding by the same process. In this way, a chain of cells may be produced. Eventually, the individual buds pinch off the parent cell and become individual yeast cells. Buds that are pinched off a hypha of a filamentous fungus behave as spores; that is, they germinate, each giving rise to a structure called a germ tube, which develops into a new hypha.

Although fragmentation, fission, and budding are methods of asexual reproduction in a number of fungi, the majority reproduce asexually by the formation of spores. Spores that are produced asexually are often termed mitospores, and such spores are produced in a variety of ways.

Sexual Reproduction

Sexual reproduction, an important source of genetic variability, allows the fungus to adapt to new environments. The process of sexual reproduction among the fungi is in many ways unique. Whereas nuclear division in other eukaryotes, such as animals, plants, and protists, involves the dissolution and re-formation of the nuclear membrane, in fungi the nuclear membrane remains intact throughout the process, although gaps in its integrity are found in some species. The nucleus of the fungus becomes pinched at its midpoint, and the diploid chromosomes are pulled apart by spindle fibres formed within the intact nucleus. The nucleolus is usually also retained and divided between the daughter cells, although it may be expelled from the nucleus, or it may be dispersed within the nucleus but detectable.

Sexual reproduction in the fungi consists of three sequential stages: plasmogamy, karyogamy, and meiosis. The diploid chromosomes are pulled apart into two daughter cells, each containing a single set of chromosomes (a haploid state). Plasmogamy, the fusion of two protoplasts (the contents of the two cells), brings together two compatible haploid nuclei. At this point, two nuclear types are present in the same cell, but the nuclei have not yet fused. Karyogamy results in the fusion of these haploid nuclei and the formation of a diploid nucleus (i.e., a nucleus containing two sets of chromosomes, one from each parent). The cell formed by karyogamy is called the zygote. In most fungi the zygote is the only cell in the entire life cycle that is diploid. The dikaryotic state that results from plasmogamy is often a prominent condition in fungi and may be prolonged over several generations. In the lower fungi, karyogamy usually follows plasmogamy almost immediately. In the more evolved fungi, however, karyogamy is separated from plasmogamy. Once karyogamy has occurred, meiosis (cell division that reduces the chromosome number to one set per cell) generally follows and restores the haploid phase. The haploid nuclei that result from meiosis are generally incorporated in spores called meiospores.

Fungi employ a variety of methods to bring together two compatible haploid nuclei (plasmogamy). Some produce specialized sex cells (gametes) that are released from differentiated sex organs called gametangia. In other fungi two gametangia come in contact, and nuclei pass from the male gametangium into the female, thus assuming the function of gametes. In still other fungi the gametangia themselves may fuse in order to bring their nuclei together. Finally, some of the most advanced fungi produce no gametangia at all; the somatic (vegetative) hyphae take over the sexual function, come in contact, fuse, and exchange nuclei.

Fungi in which a single individual bears both male and female gametangia are hermaphroditic fungi. Rarely, gametangia of different sexes are produced by separate individuals, one a male, the other a female. Such species are termed dioecious. Dioecious species usually produce sex organs only in the presence of an individual of the opposite sex.

Sexual Incompatibility

Many of the simpler fungi produce differentiated male and female organs on the same thallus but do not undergo self-fertilization because their sex organs are incompatible. Such fungi require the presence of thalli of different mating types in order for sexual fusion to take place. The simplest form of this mechanism occurs in fungi in which there are two mating types, often designated + and – (or A and a). Gametes produced by one type of thallus are compatible only with gametes produced by the other type. Such fungi are said to be heterothallic. Many fungi, however, are homothallic; i.e., sex organs produced by a single thallus are self-compatible, and a second thallus is unnecessary for sexual reproduction. Some of the most complex fungi (e.g., mushrooms) do not develop differentiated sex organs; rather, the sexual function is carried out by their somatic hyphae, which unite and bring together compatible nuclei in preparation for fusion. Homothallism and heterothallism are encountered in fungi that have not developed differentiated sex organs, as well as in fungi in which sex organs are easily distinguishable. Compatibility therefore refers to a physiological differentiation, and sex refers to a morphological (structural) one; the two phenomena, although related, are not synonymous.

Sexual Pheromones

The formation of sex organs in fungi is often induced by specific organic substances. Although called sex hormones when first discovered, these organic substances are actually sex pheromones, chemicals produced by one partner to elicit a sexual response in the other. In Allomyces (order Blastocladiales) a pheromone named sirenin, secreted by the female gametes, attracts the male gametes, which swim toward the former and fuse with them. In some simple fungi, which may have gametangia that are not differentiated structurally, a complex biochemical interplay between mating types produces trisporic acid, a pheromone that induces the formation of specialized aerial hyphae. Volatile intermediates in the trisporic acid synthetic pathway are interchanged between the tips of opposite mating aerial hyphae, causing the hyphae to grow toward each other and fuse together. In yeasts belonging to the phyla Ascomycota and Basidiomycota, the pheromones are small peptides. Several pheromone genes have been identified and characterized in filamentous ascomycetes and basidiomycetes.

Life Cycle of Fungi

In the life cycle of a sexually reproducing fungus, a haploid phase alternates with a diploid phase. The haploid phase ends with nuclear fusion, and the diploid phase begins with the formation of the zygote (the diploid cell resulting from fusion of two haploid sex cells). Meiosis (reduction division) restores the haploid number of chromosomes and initiates the haploid phase, which produces the gametes. In the majority of fungi, all structures are haploid except the zygote. Nuclear fusion takes place at the time of zygote formation, and meiosis follows immediately. Only in Allomyces and a few related genera and in some yeasts is alternation of a haploid thallus with a diploid thallus definitely known.

In the higher fungi a third condition is interspersed between the haploid and diploid phases of the life cycle. In these fungi, plasmogamy (fusion of the cellular contents of two hyphae but not of the two haploid nuclei) results in dikaryotic hyphae in which each cell contains two haploid nuclei, one from each parent. Eventually, the nuclear pair fuses to form the diploid nucleus and thus the zygote. In the Basidiomycota, binucleate cells divide successively and give rise to a binucleate mycelium, which is the main assimilative phase of the life cycle. It is the binucleate mycelium that eventually forms the basidia—the stalked fruiting bodies in which nuclear fusion and meiosis take place prior to the formation of the basidiospores.

Despite their gelatinous appearance, jelly fungi (Tremella mesenterica; also known as witch's butter) contain longitudinally septate basidia, which are formed from binucleate mycelia.

Fungi usually reproduce both sexually and asexually. The asexual cycle produces mitospores, and the sexual cycle produces meiospores. Even though both types of spores are produced by the same mycelium, they are very different in form and easily distinguished. The asexual phase usually precedes the sexual phase in the life cycle and may be repeated frequently before the sexual phase appears.

Some fungi differ from others in their lack of one or the other of the reproductive stages. For example, some fungi reproduce only sexually (except for fragmentation, which is common in most fungi), whereas others reproduce only asexually. A number of fungi exhibit the phenomenon of parasexuality, in which processes comparable to plasmogamy, karyogamy, and meiosis take place. However, these processes do not occur at a specified time or at specified points in the life cycle of the organism. As a result, parasexuality is characterized by the prevalence of heterokaryosis in a mycelium—i.e., the presence, side by side, of nuclei of different genetic composition.

Ecology of Fungi

Relatively little is known of the effects of the environment on the distribution of fungi that utilize dead organic material as food. The availability of organic food is certainly one of the factors controlling such distribution. A great number of fungi appear able to utilize most types of organic materials, such as lignin, cellulose, or other polysaccharides, which have been added to soils or waters by dead vegetation. Most saprotrophic fungi are widely distributed throughout the world, only requiring that their habitats have sufficient organic content to support their growth. However, some saprotrophs are strictly tropical and others are strictly temperate-zone forms; fungi with specific nutritional requirements are even further localized.

Fungi are found in areas that have sufficient organic material and moisture to support their growth. For example, members of the genus Armillaria are often found in forests living on trees such as hardwoods or conifers.

Moisture and temperature are two additional ecological factors that are important in determining the distribution of fungi. Laboratory studies have shown that many, perhaps the majority, of fungi are mesophilic, meaning they have an optimum growth temperature of 20–30 °C (68–86 °F). Thermophilic species are able to grow at 50 °C (122 °F) or higher but are unable to grow below 30 °C. Although the optimum temperature for growth of most fungi lies at or above 20 °C, a large number of species are able to grow close to or below 0 °C (32 °F). The so-called snow molds and the fungi that cause spoilage of refrigerated foods are examples of this group. Obviously, temperature relationships influence the distribution of various species. Certain other effects of temperature are also important factors in determining the habitats of fungi. Many coprophilous (dung-inhabiting) fungi, such as Pilobolus, although able to grow at a temperature of 20–30 °C, require a short period at 60 °C (140 °F) for their spores to germinate.

Lichens

Basic Features of Lichens

A lichen is an association between one or two fungus species and an alga or cyanobacterium (blue-green alga) that results in a form distinct from the symbionts. Although lichens appear to be single plantlike organisms, under a microscope the associations are seen to consist of millions of cells of algae (called the phycobiont) woven into a matrix formed of the filaments of the fungi (called the mycobiont). Many mycobionts are placed in a single group of Ascomycota called the Lecanoromycetes, which are characterized by an open, often button-shaped fruit called an apothecium. Although lichens had long been assumed to consist of a single fungus species and a single phycobiont, research suggests that many macrolichens also feature specific basidiomycete yeasts in the cortex of the organism. There are various types of phycobionts, though half the lichen associations contain species of Trebouxia, a single-celled green alga. There are about 15 species of cyanobacteria that act as the photobiont in lichen associations, including some members of the genera Calothrix, Gloeocapsa, and Nostoc.

Orange star lichen (Xanthoria elegans) and green lichen (Risocarpen geographica).

Authorities have not been able to establish with any certainty when and how these associations evolved, although lichens must have evolved more recently than their components and probably arose independently from different groups of fungi and algae or fungi and cyanobacteria. It seems, moreover, that the ability to form lichens can spread to new groups of fungi and algae. Lichens are a biological group lacking formal status in the taxonomic framework of living organisms. Although the mycobiont and phycobiont have Latin names, the product of their interaction, a lichen, does not. Earlier names given to lichens as a whole are considered names for the fungus alone, and much of the problem lies in the fact that the taxonomy of lichens was established before their dual nature was recognized; i.e., the association was treated as a single entity. Classification of lichens is difficult and remains controversial, though genetic analyses of the symbionts in a given lichen may serve to clarify the taxonomy of the group.

Approximately 15,000 different kinds of lichens, some of which provide forage for reindeer and products for humans, have been described. Some lichens are leafy and form beautiful rosettes on rocks and tree trunks; others are filamentous and drape the branches of trees, sometimes reaching a length of 2.75 metres (9 feet). At the opposite extreme are those smaller than a pin head and seen only with a magnifying lens. Lichens grow on almost any type of surface and can be found in almost all areas of the world. They are especially prominent in bleak, harsh regions where few plants can survive. They grow farther north and farther south and higher on mountains than most plants.

The thallus of a lichen has one of several characteristic growth forms: crustose, foliose, or fruticose. Crustose thalli, which resemble a crust closely attached to a surface, are drought-resistant and well adapted to dry climates. They prevail in deserts, Arctic and Alpine regions, and ice-free parts of Antarctica. Foliose, or leafy, thalli grow best in areas of frequent rainfall; two foliose lichens, Hydrothyria venosa and Dermatocarpon fluviatile, grow on rocks in freshwater streams of North America. Fruticose (stalked) thalli and filamentous forms prefer to utilize water in vapour form and are prevalent in humid, foggy areas such as seacoasts and mountainous regions of the tropics.

Humans have used lichens as food, as medicine, and in dyes. A versatile lichen of economic importance is Cetraria islandica, commonly called Iceland moss and sometimes used either as an appetite stimulant or as a foodstuff in reducing diets; it has also been mixed with bread and has been used to treat diabetes, nephritis, and catarrh. In general, however, lichens have little medical value. One lichen, Lecanora esculenta, is reputed to have been the manna that fell from the skies during the biblical Exodus and has served as a food source for humans and domestic animals.

Richardson's masonhalea lichen (Masonhalea richardsonii).

Lichens are well known as dye sources. Dyes derived from them have an affinity for wool and silk and are formed by decomposition of certain lichen acids and conversion of the products. One of the best-known lichen dyes is orchil, which has a purple or red-violet colour. Orchil-producing lichens include species of Ochrolechia, Roccella, and Umbilicaria. Litmus, formed from orchil, is widely used as an acid-base indicator. Synthetic coal tar dyes, however, have replaced lichen dyes in the textile industry, and orchil is limited to use as a food-colouring agent and an acid-base indicator. A few lichens (e.g., Evernia prunastri) are used in the manufacture of perfumes.

Caribou and reindeer depend on lichens for two-thirds of their food supply. In northern Canada an acre of land undisturbed by animals for 120 years or more may contain 250 kg (550 pounds) of lichens; some forage lichens that form extensive mats on the ground are Cladonia alpestris, C. mitis, C. rangiferina, and C. sylvatica. Arboreal lichens such as Alectoria, Evernia, and Usnea also are valuable as forage. An acre of mature black spruce trees in northern Canada, for example, may contain more than 270 kg (595 pounds) of lichens on branches within 3 metres (10 feet) of the ground.

Form and Function of Lichens

Although the fungal symbionts of many lichens have fruiting structures on or within their thalli and may release numerous spores that develop into fungi, indirect evidence suggests that natural unions of fungi and algae occur only rarely among some lichen groups, if indeed they occur at all. In addition, free-living potential phycobionts are not widely distributed; for example, despite repeated searches, free-living populations of Trebouxia have not been found. This paradox, an abundance of fungal spores and a lack of algae capable of forming associations, implies that the countless spores produced by lichen fungi are functionless, at least so far as propagation of the association is concerned. Some photobionts, including species of Nostoc and Trentpohlia, can exist as free-living populations, so that natural reassociations could occur in a few lichens.

Some lichens have solved or bypassed the problem of re-forming the association. In a few lichens (e.g., Endocarpon, Staurothele) algae grow among the tissues of a fruiting body and are discharged along with fungal spores; such phycobionts are called hymenial algae. When the spores germinate, the algal cells multiply and gradually form lichens with the fungus. Other lichens form structures, especially soredia, that are effective in distributing the association. A soredium, consisting of one or several algal cells enveloped by threadlike fungal filaments, or hyphae, may develop into a thallus under suitable conditions. Lichens without soredia may propagate by fragmentation of their thalli. Many lichens develop small thalloid extensions, called isidia, that also may serve in asexual propagation if broken off from the thallus.

In addition to these mechanisms for propagation, the individual symbionts have various methods of reproduction. For example, ascolichens (lichens in which the dominant mycobiont is an ascomycete) form fruits called ascocarps that are similar to those of free-living ascomycetes, except that the mycobiont's fruits are capable of producing spores for a longer period of time. The algal symbiont within the lichen thallus reproduces by the same methods as its free-living counterpart.

Most lichen phycobionts are penetrated to varying degrees by specialized fungal structures called haustoria. Trebouxia lichens have a pattern in which deeply penetrating haustoria are prevalent in associations lacking a high degree of thalloid organization. On the other hand, superficial haustoria prevail among forms with highly developed thalli. Lecanora and Lecidea, for example, have individual algal cells with as many as five haustoria that may extend to the cell centre. Alectoria and Cladonia have haustoria that do not penetrate far beyond the algal cell wall. A few phycobionts, such as Coccomyxa and Stichococcus, which are not penetrated by haustoria, have thin-walled cells that are pressed close to fungal hyphae.

The flow of nutrients and metabolites between the symbionts is the basic foundation of the symbiotic system. A simple carbohydrate formed in the algal layer eventually is excreted, taken up by the mycobiont, and transformed into a different carbohydrate. The release of carbohydrate by the phycobiont and its conversion by the mycobiont occur rapidly. Whether the fungus influences the release of carbohydrate by the alga is not known with certainty, but it is known that carbohydrate excretion by the alga decreases rapidly if it is separated from the fungus.

Carbohydrate transfer is only one aspect of the symbiotic interaction in lichens. The alga may provide the fungus with vitamins, especially biotin and thiamine, important because most lichen fungi that are grown in the absence of algae have vitamin deficiencies. The alga also may contribute a substance that causes structural changes in the fungus since it forms the typical lichen thallus only in association with an alga.

One contribution of the fungus to the symbiosis concerns absorption of water vapour from the air; the process is so effective that, at high levels of air humidity, the phycobionts of some lichens photosynthesize at near-maximum rates. The upper region of a thallus provides shade for the underlying algae, some of which are sensitive to strong light. In addition, the upper region may contain pigments or crystals that further reduce light intensity and act as filters, absorbing certain wavelengths of light.

Lichens synthesize a variety of unique organic compounds that tend to accumulate within the thallus; many of these substances are coloured and are responsible for the red, yellow, or orange colour of lichens.

A lichen thallus or composite body has one of two basic structures. In a homoiomerous thallus, the algal cells, which are distributed throughout the structure, are more numerous than those of the fungus. The more common type of thallus, a heteromerous thallus, has four distinct layers, three of which are formed by the fungus and one by the alga. The fungal layers are called upper cortex, medulla, and lower cortex. The upper cortex consists of either a few layers of tightly packed cells or hyphae that may contain pigments. A cuticle may cover the cortex. The lower cortex, which is similar in structure to the upper cortex, participates in the formation of attachment structures called rhizines. The medulla, located below the algal layer, is the widest layer of a heteromerous thallus. It has a cottony appearance and consists of interlaced hyphae. The loosely structured nature of the medulla provides it with numerous air spaces and allows it to hold large amounts of water. The algal layer, about three times as wide as a cortex, consists of tightly packed algal cells enveloped by fungal hyphae from the medulla.

A heteromerous thallus may have a stalked (fruticose), crustlike (crustose), or leafy (foliose) form; many transitional types exist. It is not known, moreover, which growth form is primitive and which is advanced. Fruticose lichens, which usually arise from a primary thallus of a different growth form (i.e., crustose, foliose), may be shrubby or pendulous or consist of upright stalks. The fruticose form usually consists of two thalloid types: the primary thallus is crustlike or lobed; the secondary thalli, which originate from the crust or lobes of the primary thallus, consist of stalks that may be simple, cup-shaped, intricately branched, and capped with brown or red fruiting bodies called apothecia. Fruticose forms such as Usnea may have elongated stalks with a central solid core that provides strength and elasticity to the thallus.

The crustose thallus is in such intimate contact with the surface to which it is attached that it usually cannot be removed intact. Some crustose lichens grow beneath the surface of bark or rock so that only their fruiting structures penetrate the surface. Crustose lichens may have a hypothallus—i.e., an algal-free mat of hyphae extending beyond the margin of the regular thallus. Crustose form varies: granular types such as Lepraria, for example, have no organized thalloid structure; but some Lecanora species have highly organized thalli, with lobes that resemble foliose lichens lacking a lower cortex.

The foliose forms are flat, leaflike, and loosely attached to a surface. The largest known lichens have a foliose form; species of Sticta may attain a diameter of about a metre. Other common foliose genera include Cetraria, Parmelia, Peltigera, and Physcia. Umbilicaria, called the common rock tripe, differs from other foliose forms in its mode of attachment in that its platelike thallus attaches at the centre to a rock surface.

Dog lichen (Peltigera canina) has a foliose thallus that grows in a flat layer with rounded lobes. It is often found on surfaces such as walls and sand dunes.

The complex fruiting bodies (ascocarps) of lichen fungi are of several types. The factors that induce fruiting in lichens have not been established with certainty. Spores of lichen fungi (ascospores) are of extremely varying sizes and shapes; e.g., Pertusaria has one or two large spores in one ascus (saclike bodies containing the ascospores), and Acarospora may have several hundred small spores per ascus. Although in most species the ascospore generally has one nucleus, it may be single-celled or multicellular, brown or colourless; the Pertusaria spore, however, is a single cell containing 200 nuclei. Another type of fungal spore may be what are sometimes called spermatia (male fungal sex cells) or pycnidiospores; it is not certain that these structures have the ability to germinate and develop into a fungal colony. Few lichen fungi produce conidia, a type of asexual spore common among ascomycetes.

The metabolic activity of lichens is greatly influenced by the water content of the thallus. The rate of photosynthesis may be greatest when the amount of water in the thallus is from 65 to 90 percent of the maximum. During drying conditions, the photosynthetic rate decreases; below 30 percent it is no longer measurable. Although respiration also decreases rapidly below 80 percent water content, it persists at low rates even when the thallus is air-dried. Since lichens have no mechanisms for water retention or uptake from the surface to which they are attached, they very quickly lose the water vapour they absorb from the air. The rapid drying of lichens is a protective device; i.e., a moisture-free lichen is more resistant to temperature and light extremes than is a wet one. Frequent drying and wetting of a thallus is one of the reasons lichens have a slow growth rate.

Maximum photosynthesis in lichens takes place at temperatures of 15–20 °C (59–68 °F). More light is needed in the spring and summer than in the winter. The photosynthetic apparatus of lichens is remarkably resistant to cold temperatures. Even at temperatures below 0 °C (32 °F), many lichens can absorb and fix considerable amounts of carbon dioxide. Respiration is much less at low temperatures so that, in nature, the winter months may be the most productive ones for lichens.

Outline of Classification of Fungi

Since the 1990s, dramatic changes have occurred in the classification of fungi. Improved understanding of relationships of fungi traditionally placed in the phyla Chytridiomycota and Zygomycota has resulted in the dissolution of outmoded taxons and the generation of new taxons. The Chytridiomycota is retained but in a restricted sense. One of Chytridiomycota's traditional orders, the Blastocladiales, has been raised to phylum status as the Blastocladiomycota. Similarly, the group of anaerobic rumen chytrids, previously known as order Neocallimastigales, has been recognized as a distinct phylum, the Neocallimastigomycota. The phylum Zygomycota is not accepted in the phylogenetic classification of fungi because of remaining doubts about relationships between the groups that have traditionally been placed in this phylum. The consequences of this decision are the recognition of the phylum Glomeromycota and of four subphyla incertae sedis (Latin for "of uncertain position"): Mucoromycotina, Kickxellomycotina, Zoopagomycotina, and Entomophthoromycotina.

Saddle fungi (Helvella infula) are members of the phylum Ascomycota, one of the seven phyla that compose the kingdom Fungi.

The true fungi, which make up the monophyletic clade called kingdom Fungi, comprise seven phyla: Chytridiomycota, Blastocladiomycota, Neocallimastigomycota, Microsporidia, Glomeromycota, Ascomycota, and Basidiomycota (the latter two being combined in the subkingdom Dikarya). The group of ancestral fungi is thought to be represented by the present-day Chytridiomycota, although the Microsporidia may be an equally ancient sister group. The first major steps in the evolution of higher fungi were the loss of the chytrid flagellum and the development of branching, aseptate fungal filaments, which occurred as terrestrial fungi diverged from water molds 600 million to

800 million years ago. Septate filaments evolved as the Glomeromycota diverged from a combined clade of pre-basidiomycota and pre-ascomycota fungi about 500 million years ago. Hyphae with the characteristic appearance of modern Basidiomycota can be seen in some of the earliest known specimens of plant fossils. Therefore, Ascomycota and Basidiomycota probably diverged as so-called sister groups, which are placed together in subkingdom Dikarya, about 300 million years ago. The easily recognizable mushroom fungi probably diversified 130 million to 200 million years ago, soon after flowering plants became an important part of the flora and well before the age of dinosaurs. A relatively recent evolutionary radiation, perhaps 60 million to 80 million years ago, of anaerobic Chytridiomycota occurred as grasses and grazing mammals became more abundant; the chytrid fungi serve as symbionts within the rumen of such animals, thereby enabling the grazing mammals to digest grasses.

Classification of the Fungi

Distinguishing Taxonomic Features

The following classification is adapted from Ainsworth & Bisby's Dictionary of the Fungi, 10th ed., and has been amended to adopt the phylogenetic arrangement from the Assembling the Fungal Tree of Life (AFTOL) project funded by the U.S. National Science Foundation. AFTOL is a work in progress, and uncertainties remain about the exact relationships of many groups. These uncertain groups are indicated in the annotated classification below by the term incertae sedis, meaning "of uncertain position," the standard term for a taxonomic group of unknown or undefined relationship. The phylogenetic classification of fungi divides the kingdom into 7 phyla, 10 subphyla, 35 classes, 12 subclasses, and 129 orders.

PROTOZOA

Protozoan is an organism, usually single-celled and heterotrophic (using organic carbon as a source of energy), belonging to any of the major lineages of protists and, like most protists, typically microscopic. All protozoans are eukaryotes and therefore possess a "true," or membrane-bound, nucleus. They also are nonfilamentous (in contrast to organisms such as molds, a group of fungi, which have filaments called hyphae) and are confined to moist or aquatic habitats, being ubiquitous in such environments worldwide, from the South Pole to the North Pole. Many are symbionts of other organisms, and some species are parasites.

Modern ultrastructural, biochemical, and genetic evidence has rendered the term protozoan highly problematic. For example, protozoan historically referred to a protist that has animal-like traits, such as the ability to move through water as though "swimming" like an animal. Protozoans traditionally were thought to be the progenitors of modern animals, but contemporary evidence has revealed that this is not the case for most protozoans. In fact, modern science has shown that the protozoans represent a very complicated grouping of organisms that do not necessarily share a common evolutionary history. This unrelated, or paraphyletic, nature of the protozoans has caused scientists to abandon the term protozoan in formal classification schemes. Hence, the subkingdom Protozoa is now considered obsolete. Today the term protozoan is used informally in reference to nonfilamentous heterotrophic protists.

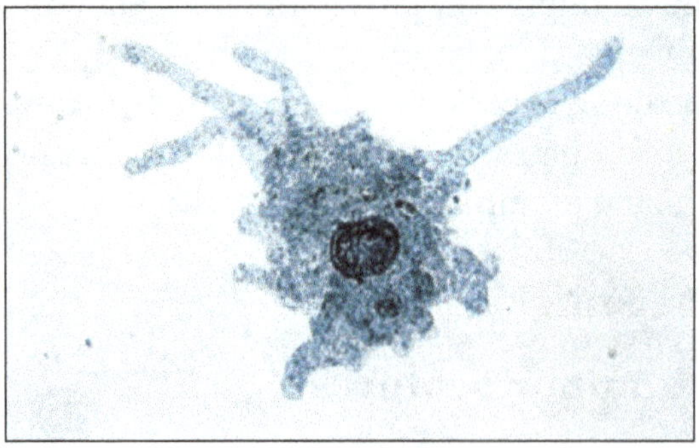

Amoeba (magnified).

Commonly known protozoans include representative dinoflagellates, amoebas, paramecia, and the malaria-causing Plasmodium.

Features of Protozoans

Although protozoans are no longer recognized as a formal group in current biological classification systems, protozoan can still be useful as a strictly descriptive term. The protozoans are unified by their heterotrophic mode of nutrition, meaning that these organisms acquire carbon in reduced form from their surrounding environment. However, this is not a unique feature of protozoans. Furthermore, this description is not as straightforward as it seems. For instance, many protists are mixotrophs, capable of both heterotrophy (secondary energy derivation through the consumption of other organisms) and autotrophy (primary energy derivation, such as through the capture of sunlight or metabolism of chemicals in the environment). Examples of protozoan mixotrophs include many chrysophytes. Some protozoans, such as Paramecium bursaria, have developed symbiotic relationships with eukaryotic algae, while the amoeba Paulinella chromatophora remarkably appears to have acquired autotrophy via relatively recent endosymbiosis of a cyanobacterium (a blue-green alga). Hence, many protozoans either perform photosynthesis themselves or benefit from the photosynthetic capabilities of other organisms. Some algal species of protozoans, however, have lost the ability to photosynthesize (e.g., Polytomella species and many dinoflagellates), further complicating the concept of "protozoan".

Protozoans are motile; nearly all possess flagella, cilia, or pseudopodia that allow them to navigate their aqueous habitats. However, this commonality does not represent a unique trait among protozoans; for example, organisms that are clearly not protozoans also produce flagella at various stages in their life cycles (e.g., most brown algae). Protozoans are also strictly non-multicellular and exist as either solitary cells or cell colonies. Nevertheless, some colonial organisms (e.g., Dictyostelium discoideum, supergroup Amoebozoa) exhibit high levels of cell specialization that border on multicellularity.

The descriptive guidelines presented above exclude many organisms, such as flagellated photosynthetic taxa (formerly Phytomastigophora), that were considered protozoans by older classification schemes. Organisms that fit the contemporary definition of a protozoan are found in all major groups of protists that are recognized by protistologists, reflecting the paraphyletic nature of protozoans.

The most important groups of free-living protozoans are found within several major evolutionary clusters of protists, including the ciliates (supergroup Chromalveolata), the lobose amoebae (supergroup Amoebozoa), the filose amoebae (supergroup Rhizaria), the cryptomonads (supergroup Chromalveolata), the excavates (supergroup Excavata), the opisthokonts (supergroup Opisthokonta), and the euglenids (Euglenozoa). These groups of organisms are important ecologically for their role in microbial nutrient cycles and are found in a wide variety of environments, from terrestrial soils to freshwater and marine habitats to aquatic sediments and sea ice. Significant protozoan parasites include representatives from Apicomplexa (supergroup Chromalveolata) and the trypanosomes (Euglenozoa). Organisms from these groups are the causative agents of human diseases such as malaria and African sleeping sickness. Owing to the prevalence of these human pathogens, and to the ecological importance of the free-living protozoan groups mentioned above, much is known about these groups. This article therefore concentrates on the biology of these comparatively well-characterized protozoans. At the end of this article is a summary of the contemporary protistan classification scheme.

Size Range and Diversity of Structure

Protozoans range in diameter from a few thousandths of a millimetre to several millimetres. Because the group contains many unrelated or loosely related organisms, enormous diversity in structure and form exists.

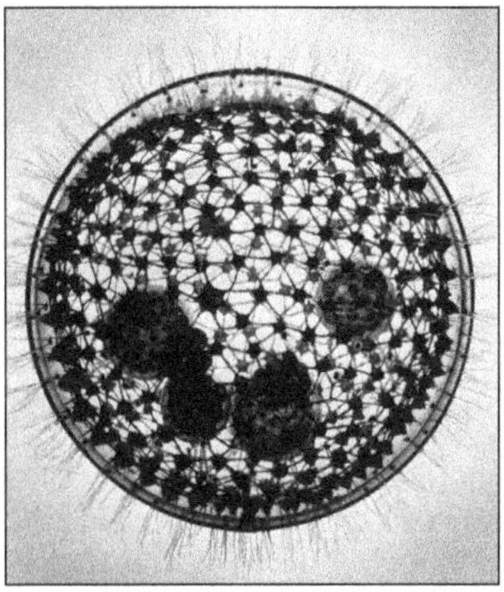

Glass model of protozoan colony Volvox (magnified about 40×).
Courtesy of the American Museum of Natural History, New York City.

Flagellated Protozoans

The flagellated protozoans range from a simple oval cell with one or more flagella to the structural sophistication of the collared flagellates (choanoflagellates, supergroup Opisthokonta). The collared flagellates lack photosynthetic pigments and are therefore colourless. They have a single flagellum surrounded by a delicate circular collar of fine pseudopodia (microvilli) on which they trap food particles. In some marine species the whole cell is enclosed in an elaborate, open latticelike basket formed from strands of silica. Although some dinoflagellates (supergroup Chromalveolata) still contain plant pigments and rely to a greater or lesser degree on photosynthesis, many members have lost the ability to photosynthesize. All dinoflagellates are surrounded by a cell wall armour with a complicated pattern and possess two flagella, one of which beats in a transverse plane around the equator of the cell while the other beats in a longitudinal plane.

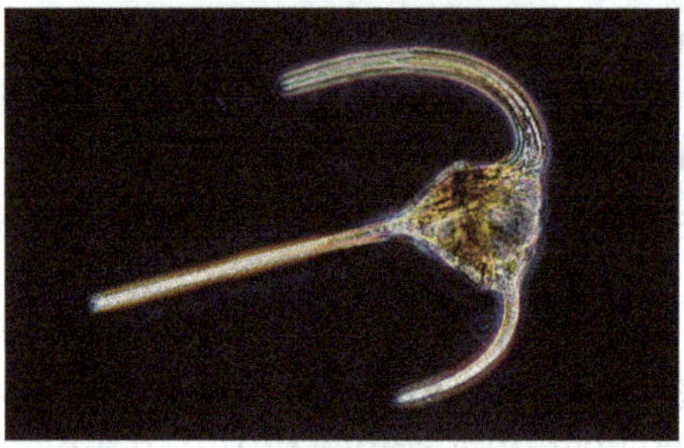

Many other flagellated protozoans can develop stalks that connect them to a substrate, either as single cells (e.g., the genus Paraphysomonas) or as colonies (e.g., the genus Codosiga). Other flagellated taxa may exist as swimming colonies; for example, in Sphaeroeca and Spongomonas many individual flagellated organisms are embedded in an unstalked gelatinous sphere.

Amoebae and Pseudopodia

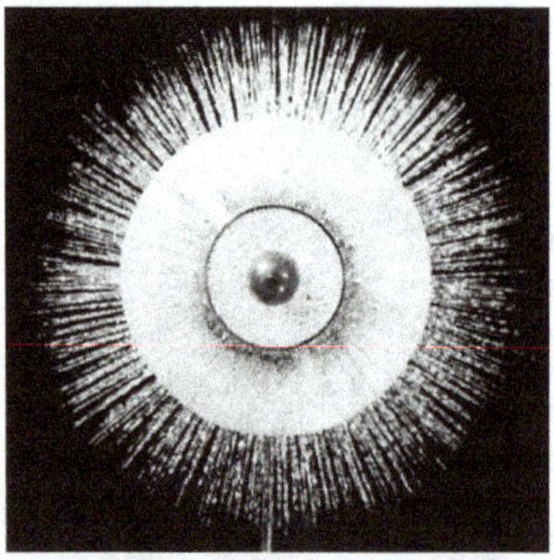

Radiolarian (glass model) Courtesy of the American Museum of Natural History, New York City.

The amoebae also are extremely diverse. Amoebae are defined based on pseudopodia type: those with thin, or filose, pseudopods, which may be reinforced by stiff microtubule proteins, are classified in the supergroup Rhizaria (e.g., foraminiferans and radiolarians), whereas those with lobose pseudopods, which are blunt and are not reinforced, are classified in the supergroup Amoebozoa. Both groups of amoebae can be "naked" or housed inside a shell, or test, composed of organic or inorganic materials.

The naked amoebae are the simplest of the amoebae. They have no defined shape and extend one or many lobose pseudopodia. Many of these lobose amoebae, including those in the genera Mastigamoeba and Mastigella, also possess flagella in the vegetative (resting) phase. At the opposite extreme are the complex foraminiferans, which live inside multichambered calcareous shells up to several millimetres in diameter. The filose pseudopodia of foraminiferans are known as reticulopodia and extend from the aperture of the largest chamber of the shell, forming a complicated, sticky branching network. Rhizarian amoebae that are known commonly as radiolarians form shells from silica or strontium sulfate; in some the shell has so many holes that the structure resembles a sponge. The polyphyletic heliozoans, or sun protozoans, have radiating pseudopodia (axopodia) that extend like spokes from the central body; microtubules support an outer layer of cytoplasm. Many heliozoans are members of Rhizaria; however, some are placed in Chromalveolata.

Ciliated Protozoans

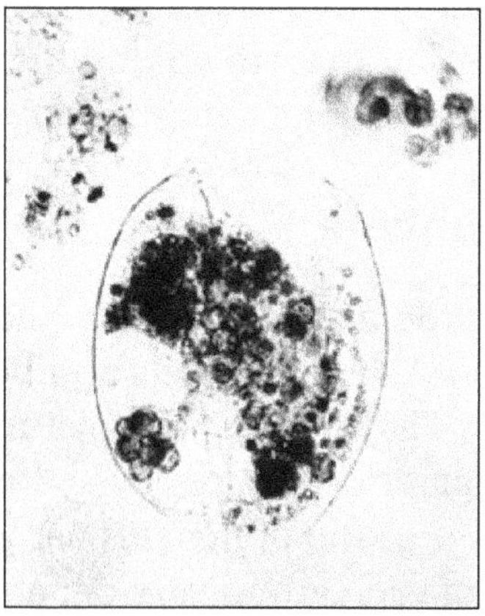

Euplotes patella digesting algae. Movement of cilia has started a whirlpool of algae at right.

The ciliates are the most structurally homogeneous group, although even they have evolved considerable variation on the cilia-covered cell. In some species (e.g., the hypotrich Euplotes) the cilia are combined to form thick conical structures, called cirri, which the ciliate uses to crawl along surfaces, rather like small limbs. In other species the cilia virtually disappear from the main body of the cell, but the circle of cilia around the mouth becomes well developed (as in the oligotrich Strombidium and the tintinnid ciliates). The peritrich ciliates have developed stalks and attach to plants and animals as a means of dispersal. Many peritrichs (e.g., Epistylis) form branching colonies.

The suctorian ciliates have completely lost their cilia in the adult phase. They have instead developed a stalk and many tentacles, which they use to capture passing prey, usually other ciliates. Because they cannot swim, they produce motile ciliated offspring, which settle elsewhere and then transform into the feeding stage, thus avoiding overcrowding.

Parasitic Protozoans

Although the parasitic protozoans tend to be less structurally complex than free-living forms, considerable variation may occur during the course of their life cycles. Plasmodium, the malarial parasite that lives inside the liver and red blood cells of humans and the gut of its insect vector (the Anopheles mosquito), undergoes various changes in form through its asexual and sexual phases of development. Among the parasitic flagellates, the trypanosomes and their relatives (kinetoplastids), morphological variation occurs during the various stages of the life cycle in both the mammalian and insect hosts. Among species of Leishmania, which cause visceral leishmaniasis (kala-azar), cutaneous leishmaniasis (Oriental sore), and mucocutaneous leishmaniasis (espundia), two distinctly different forms occur. Rounded, nonflagellated forms called amastigotes feed and divide inside macrophage cells in different regions of the human body, while in the gut of the insect vector there occurs a flagellated form called a promastigote. Members of the genus Trypanosoma, which cause sleeping sickness and other diseases, have flagellated forms with different morphologies. At some stage in the life cycle, all assume the trypomastigote form—i.e., slender with part of the flagellum running over the body and attached to it by a finlike extension to form an undulating membrane. They may also occur as amastigote (stumpy flagella) or promastigote forms.

Distribution and Abundance

Protozoans have colonized a wide array of aquatic and terrestrial habitats from the Arctic and Antarctic to equatorial zones. In soils and bogs, protozoans form part of a complex microbial community. They live in the moisture films surrounding soil particles, so that they are actually aquatic organisms, even though living in a terrestrial environment. Between 10,000 and 100,000 organisms per gram of soil may inhabit fertile land; the relative proportions of each group vary depending on soil type and latitude. In Antarctic soils flagellates and testate (shell-dwelling) amoebae predominate, while in temperate woodland soils ciliates are more numerous.

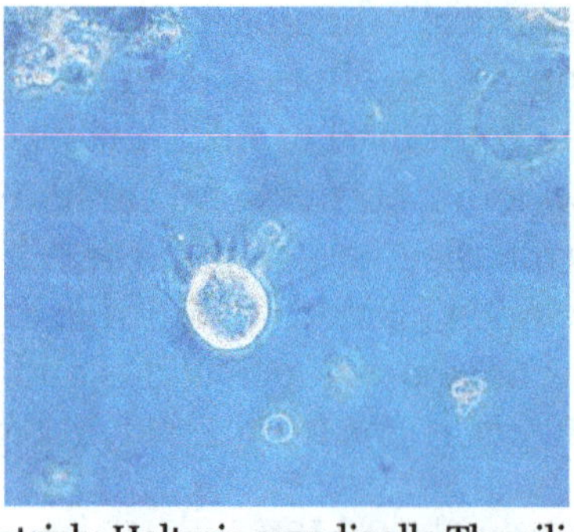

oligotrich; Halteria grandinella The ciliated oligotrich Halteria grandinella.

In the open waters of lakes, estuaries, and the ocean, protozoans form an important component of the floating (planktonic) community. They are often present in densities of tens of thousands per litre of water. Most planktonic protozoa feed on bacteria, algae, other protozoans, and small animals. The most common planktonic protozoans include a variety of flagellated taxa, ciliates—especially oligotrichs and tintinnids (which live inside small tubes, or loricae)—and the exclusively marine foraminiferans and radiolarians. Foraminiferans have been found at depths of 4,000 metres (about 13,120 feet), and some protozoans have been observed around hydrothermal vents on the ocean floor.

Ecological and Industrial Importance of Protozoans

Protozoans play important roles in the fertility of soils. By grazing on soil bacteria, they regulate bacterial populations and maintain them in a state of physiological youth—i.e., in the active growing phase. This enhances the rates at which bacteria decompose dead organic matter. Protozoans also excrete nitrogen and phosphorus, in the form of ammonium and orthophosphate, as products of their metabolism, and studies have shown that the presence of protozoans in soils enhances plant growth.

Protozoans play important roles in wastewater treatment processes, in both activated sludge and slow percolating filter plants. In both processes, after solid wastes are removed from the sewage, the remaining liquid is mixed with the final sludge product, aerated, and oxidized by aerobic microorganisms to consume the organic wastes suspended in the fluid. In the activated sludge process, aerobic ciliates consume aerobic bacteria, which have flocculated (formed loose aggregates, making them easily separated from liquid). In the percolating filter process, substrates are steeped in microorganisms, such as fungi, algae, and bacteria, which provide food for oxidizing protozoans. In the final stages of both processes, solids settle out of the cleaned effluent in the settlement tank. Treatment plants with no ciliates and only small numbers of amoebae and flagellates produce turbid effluents containing high levels of bacteria and suspended solids. Good-quality, clean effluents are produced in the presence of large ciliated protozoan communities because they graze voraciously on dispersed bacteria and because they have the ability to flocculate suspended particulate matter and bacteria.

Protozoans probably play a similar role in polluted natural ecosystems. Indeed, there is evidence that they, by feeding on oil-degrading bacteria, increase bacterial growth in much the same way that they enhance rates of decomposition in soils, thereby speeding up the breakdown of oil spillages.

Some radiolarians and foraminiferans harbour symbiotic algae that provide their protozoan hosts with a portion of the products of photosynthesis. The protozoans reciprocate by providing shelter and carbon and essential phytonutrients. Many ciliates contain endosymbiotic algae, and one species, Mesodinium rubrum, has formed such a successful relationship with its red-pigmented algal symbiont that it has lost the ability to feed and relies entirely on symbiosis for its livelihood. Mesodinium often forms dense red blooms, or red tides, when it reaches high densities in water. Among the ciliates with endosymbionts, Mesodinium is the only completely photosynthetic species. Other ciliates achieve photosynthesis in another way. Although they do not have symbiotic algae, they consume plantlike flagellates, sequester the organelles that contain the plant pigments, and use them for photosynthesis. These organelles are known as plastids. Because the isolated plastids eventually age and die, they must be replaced continuously.

The impact of protozoan grazing on phytoplankton can be considerable. It has been estimated that at least half of the phytoplankton production in marine waters is consumed by protozoans. Like the soil protozoans, these planktonic protozoans excrete nitrogen and phosphorus at high rates. The protozoans are a fundamental component in recycling essential nutrients (nitrogen and phosphorus) to the phytoplankton.

Protozoans and Disease

Parasitic protozoans have invaded and successfully established themselves in hosts from practically every animal phylum. The best-studied parasitic species are those of medical and agricultural relevance. The trypanosomes, for example, cause a number of important diseases in humans. African sleeping sickness is produced by two subspecies of Trypanosoma brucei—namely, T. brucei gambiense and T. brucei rhodesiense. The life cycle of T. brucei has two hosts: a human (or other mammal) and the bloodsucking tsetse fly, which transmits the parasite between humans.

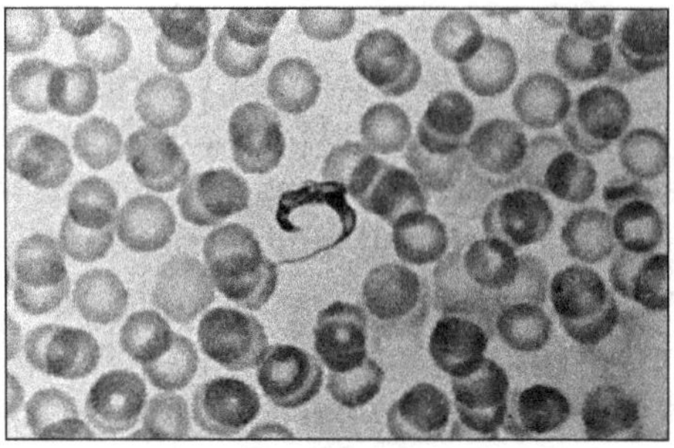

Trypanosome with human red blood cells (highly magnified).

Trypanosomes live in the blood plasma and the central nervous system of humans and have evolved an ingenious way of fooling the immune system of the host. Upon contact with a parasite, the immune system generates antibodies that recognize the specific chemical and physical nature of the parasite and actively neutralize it. Just as the host's immune system is beginning to win the battle against the parasite and the bulk of the population is being recognized and destroyed by host antibodies, the parasite is able to shed its glycoprotein coat, which is attached to the cell surface, and replace it with a coat containing different amino acid sequences. Thus, the parasite essentially changes its makeup. These alternate forms are known as antigenic variants, and it has been estimated that each species may have as many as 100 to 1,000 such variants. The host must produce a new set of antibodies against each new variant, and in the meantime the parasite has time to replenish its numbers. Ultimately, unless the disease is treated, the parasite wins the battle and the host dies. Such antigenic variation makes the development of an effective vaccine against certain parasitic protozoan diseases virtually impossible.

A close relative of T. brucei, Trypanosoma cruzi, causes Chagas disease, or American trypanosomiasis. Vector hosts include bugs of the genus Rhodnius and other arthropods, such as lice and bedbugs. In humans the nonflagellated (amastigote) form of the parasite lives inside macrophage cells, the cells of the central nervous system, and muscle tissue, including the heart, where it grows and divides. Short trypomastigote flagellated forms periodically appear in the blood, where they are readily taken up by the bloodsucking vector hosts. These flagellated forms do not divide in the blood; reproduction occurs only in the amastigote intracellular forms.

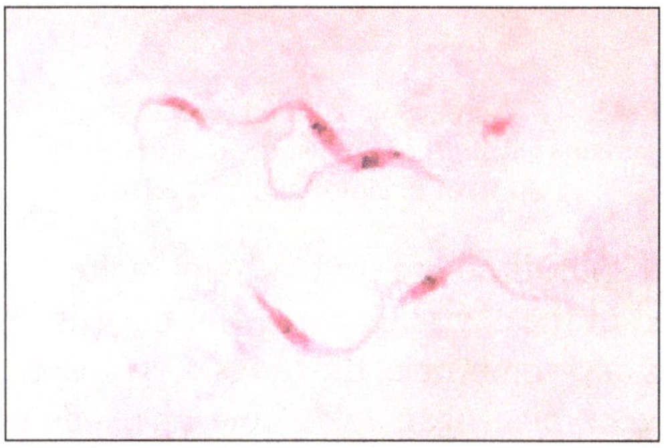

Trypanosoma cruzi Photomicrograph of Trypanosoma cruzi, the causative agent of Chagas disease.

Relatives of the trypanosomes, species of the genus Leishmania, cause a variety of diseases worldwide, known as leishmaniasis. Like T. cruzi, these are intracellular parasites of the macrophage cells. The intermediate, or vector, hosts are a variety of sand fly species (subfamily Phlebotominae). In cutaneous leishmaniasis the infected macrophages remain localized at the site of the infection, causing an unsightly lesion, but in visceral leishmaniasis the infected macrophages are carried by the blood to the visceral organs. This latter disease is characterized by enlargement of the spleen and liver, leading to the distended abdomen that is typical of kala-azar. In mucocutaneous leishmaniasis the initial skin infection spreads to the mucous membranes of the face (the nose, mouth, and throat), producing a lesion that can cause destruction of part of the face.

Malaria, which is caused by the apicomplexan protozoan Plasmodium, remains a serious disease despite measures that can be taken to control and eradicate the mosquito vector host and despite the availability of an array of antimalarial drugs. The life cycle is fundamentally identical among the five species of Plasmodium, but the pathology of the disease varies in the frequency and severity of attacks and in the occurrence of relapses. Problems in controlling the disease include the development of resistance to insecticides by the mosquito and the evolution of drug resistance by the parasite.

Prophylactic drugs taken before and during a visit to areas where malaria is endemic may prevent the disease from forming in persons who have no natural resistance. Antigenic variation does not appear to occur in Plasmodium, which is promising for vaccine development.

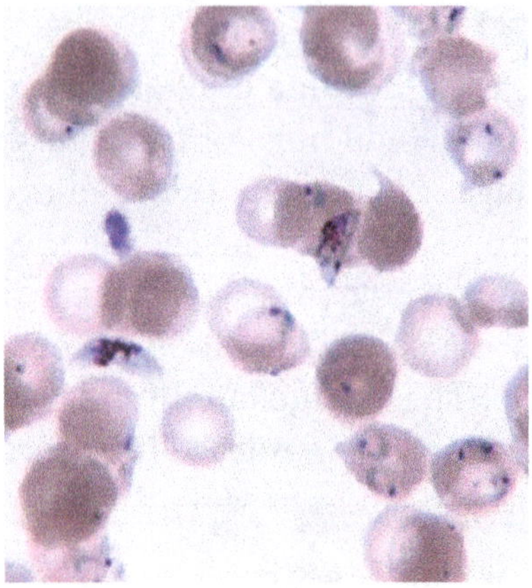

Blackwater fever Micrograph of blood cells showing ring-forms (circular organisms within the cells) and gametocytes (purplish oblong shapes) of Plasmodium falciparum.

The apicomplexan Cryptosporidium is a protozoan parasite of humans and other mammals that was discovered in the 1970s. It has a one-host life cycle and lives inside the cells lining the intestines and sometimes the lungs. Cryptosporidium carries out all the asexual reproductive stages typical of an apicomplexan inside a single host and is passed from host to host in a resistant cyst stage called an oocyst. The disease caused by the parasite is typified by severe diarrhea and vomiting. Although there is no drug treatment, most healthy people recover quickly. In persons who have impaired immune systems, such as AIDS patients, however, Cryptosporidium can cause serious infections.

Form and Function

The Protozoan Cell

The protozoan cell carries out all of the processes—including feeding, growth, reproduction, excretion, and movement—necessary to sustain and propagate life. The cell is enclosed in a membrane called the plasma membrane. Like all membranous structures in the eukaryotic cell, the plasma membrane is composed of mostly lipid and some protein molecules. The plasma membrane is a barrier between the cell cytoplasm and the outside liquid environment. Some substances, such as oxygen, readily pass through the membrane by diffusion (passive transport), while others must be transported across at the expense of energy (active transport). Cilia and flagella arising from the cell are also sheathed in the cell membrane; this is in contrast to bacterial flagella, which are not surrounded by a membrane.

The cell also has internal membranes, which are not as thick as the plasma membrane.

Among these are the endoplasmic reticulum, whose membranes separate compartments of the cell, thereby allowing different conditions to be maintained in various parts—e.g., separation of deleteriously reactive substances. Enzymes are arranged on the surface of the endoplasmic reticulum; one such enzyme system catalyzes the activity of the ribosomes during protein synthesis. The Golgi apparatus is a cluster of flattened vesicles, or cisternae, associated with the endoplasmic reticulum. The vesicles are involved in membrane maturation and the formation and storage of the products of cell synthesis, as in the formation of scales on the surface coat of some flagellates, for example. The scales are formed within the Golgi and are transported by the vesicles to the plasma membrane, where they are incorporated onto the surface of the cell. The Golgi apparatus is poorly evident in most ciliates and absent from some amoebae.

All protozoans possess at least one nucleus, and many species are multinucleate. The genetic material DNA (deoxyribonucleic acid) is contained within the chromosomes of the nucleus. Each nucleus is bounded by two unit membranes possessing pores that permit the passage of molecules between the cytoplasm and the nucleoplasm. Most ciliates have two types of nuclei: micronuclei and macronuclei. The macronucleus is the somatic, or nonreproductive, nucleus. It is large and it is polyploid, meaning that it contains more than two sets of chromosomes (the condition of two sets of chromosomes is described as diploid). In contrast, the micronucleus is germinal (responsible for transfer of genetic information during sexual reproduction) and diploid. The macronucleus can be quite variable in shape, resembling in some species a string of beads or a horseshoe. It directs the normal functioning of the cell and usually disintegrates during sexual reproduction, to be re-formed from the products of micronuclear division after the sexual phase is completed.

Almost all protozoans contain double-membrane mitochondria; the inner membrane forms flattened, tubular, or discoidal extensions (cristae) into the mitochondrial interior in order to increase the surface area of the respiratory machinery, and the outer membrane forms the boundary of the organelle. Mitochondria are the sites of cellular respiration in most eukaryotes. Species that do not require oxygen (anaerobes), such as those that live in the intestinal tract of their hosts or those that occupy special anaerobic ecological niches, lack mitochondria. Instead, they have energy-generating organelles, such as hydrogenosomes and mitosomes, that belong to the family of organelles called microbodies. These oblong or spherical membrane-bound organelles, about 1–2 micrometres (μm; 1 micrometre = 3.9 × 10–5 inch) in length, are believed to be the site of fermentative processes. They contain enzymes that oxidize pyruvate to acetate and carbon dioxide, resulting in the release of hydrogen sulfide under anaerobic conditions.

Organisms that live in a liquid environment with a lower concentration of ions than is found in the interior of their cells—an osmotically hypotonic environment—gradually gain water if they equilibrate with their habitat. If this process remains unchecked, the cell swells and bursts. In protozoans the maintenance of the osmotic gradient between the cell cytoplasm and the environment is achieved by the contractile vacuole. These membrane-bound organelles are situated close to the plasma membrane. They swell with water periodically and then suddenly contract and disappear, forcing their contents from the cell in repeated cycles. In some amoebae and some flagellated taxa the contractile vacuole is formed when smaller vesicles combine with the main vacuole. In the ciliates the contractile vacuole is fed by a complex system of feeder canals, which are in turn fed by a complex network of vesicles and fine tubules within the cytoplasm.

Protozoans have transitory food or digestive vacuoles. The number of these membrane-bound cell organelles depends on the feeding habits of the organism. Some species may have many, whereas others may contain only one or two at any one time. In ciliates the food vacuoles form at the base of the cytopharynx, whereas in species without a cell "mouth," or cytostome, the vacuoles form near the cell membrane at the site where food is ingested.

Within the cell, structural proteins of various types form the cytoskeleton (cell skeleton) and the locomotory appendages. They include microfilaments formed of a contractile protein also found in the muscles of animals (actin) and cylindrical microtubules formed from filaments of the protein tubulin. Microtubules are particularly important in the structural formation and functioning of cilia and flagella. Filopodia of certain rhizarian species are supported by microtubules.

Characteristics of Locomotion

Protozoans exhibit diverse modes of locomotion across the various groups, but the modes of locomotion can be broadly divided into flagellar, ciliary, and amoeboid movement. Only the ciliates among the three major motility groups of protozoans, however, represent a truly monophyletic group (or single evolutionary line). (Some non-ciliates, such as those of group Opalinata, possess cilia-like organelles that are fundamentally different from true cilia.) In contrast, flagella and pseudopodia are present in a wide variety of distantly related taxa.

Flagellar Propulsion

Flagellar propulsion is employed during some stages in the life cycles of certain amoebae, including the vegetative phase of some genera, such as Mastigamoeba and Mastigella. The eukaryotic flagellum is a membrane-bound, whiplike structure found not only in protozoans but in animals as well (such as in sperm, the male reproductive cells of animals). The structure of the eukaryotic flagellum consists of a cylinder (axoneme) made up of a pair of central microtubules surrounded and joined by cross-bridges to a circle of nine pairs of microtubules. This "nine-plus-two" arrangement of the microtubules in the axoneme is surrounded by cytoplasm and ensheathed in cell membrane. The flagellum arises from the basal body, or kinetosome, within the cell.

The undulating motion of the flagellum is normally generated at its base. The waves move along the flagellum to produce a force on the water acting along the long axis of the organelle in the direction of the wave. The speed of movement is determined by the length of the flagellum and by the size of, and distance between, the waves it generates. Species of monophyletic stramenopiles (heterokonts) have tripartate tubular hairs (mastigonemes) arising at right angles to the flagellum along its length, whereas other groups, such as the dinoflagellates and euglenids, have slender, simpler hairs called flimmer filaments. Either structure improves the effectiveness of the flagellar stroke, altering the movement of water produced by undulations of the flagellum by reversing its flow toward the flagellar base. Swimming speeds achieved by flagella are relatively low.

Cilium Structure and Beat

Ciliates have an increased number of beating flagella on the cell surface, thereby enabling greater power and speeds to be developed against viscous forces. The structure of a cilium is identical to that of a flagellum, but the cilium is considerably shorter. Cilia are a type of flagella arranged in closely aligned longitudinal rows called kineties. A complex system of fibres and microtubules arising from the basal bodies, or kinetosomes, of each cilium connects it to its neighbouring cilia in the kinety and to adjacent ciliary rows. In some species the body cilia may be reduced to specialized cirri, where the kinetosomes are not arranged in the usual rows but instead have a hexagonal pattern interlinked at several levels by fibres and microtubules.

The effective stroke of the cilium is usually planar, but in the recovery stroke the cilium sweeps out to the side, creating an overall beat with a three-dimensional pattern. The cilium performs work against the viscous force of the water during both the effective and the recovery strokes. To be effective, each cilium must beat in a coordinated manner with its neighbouring cilia. A synchronized beat is passed along a ciliary row by means of a hydrodynamic linkage between the cilia. During a beat, each cilium displaces a layer of surrounding water. Displaced water layers overlap between cilia and, as a consequence, interference occurs between the movements of adjacent cilia, creating a hydrodynamic linkage.

Amoeboid Movement

Amoeboid movement is achieved by pseudopodia and involves the flow of cytoplasm as extensions of the organism. The process is visible under the light microscope as a movement of granules within the organism. The basic locomotory organelle is the pseudopodium. The way in which movement is effected can vary slightly among groups but generally involves the polymerization of cytoskeletal proteins (actin and myosin) at the leading edge of the pseudopod, followed by the flow of cytoplasmic material into the vacancy produced through the polymerization process. The flow of cytoplasm provides the momentum necessary to propel the organism further in its direction of movement. Additional forces driving the amoeboid movement involve the "eupodium," which extends into a potential substrate for a grab-like traction, similar to a tank tread. Pushing force is also generated in the posterior end of the organism by contractions of the cytoskeletal proteins.

A variety of pseudopodial types are found among the naked and testate amoebae. In some species a single pseudopodium is extended at any one time; in others, numerous tubular pseudopodia are extended simultaneously. Some amoebae appear saclike throughout locomotion, and no pseudopodia are obvious. The numerous long, stiff protoplasmic extensions (axopodia) of the amoeboid (paraphyletic) heliozoans shorten and lengthen—the forward axopodia lengthen and become attached, while the posterior axopodia detach and retract—and the amoeba rolls slowly along. The foraminiferans move by extending slender pseudopodia (filapodia), which may be several millimetres long in some species. The extending filopodia branch and fuse with each other so that there is a complex, continuously changing network of pseudopodia pulling the organism along.

Respiration and Other Energy-generating Pathways

Aerobic Protozoans

Most species of free-living protozoans appear to be obligate aerobes (they cannot survive without oxygen). As in the cells of animals, plants, and fungi, their respiration is based on oxidation (with molecular oxygen, O2) of the six-carbon glucose molecule, resulting in the formation of carbon dioxide molecules and water. In protozoans and eukaryotes in general, metabolism and respiration occur stepwise via three specific pathways: the Embden-Meyerhof-Parnas pathway (glycolysis), the tricarboxylic acid cycle (also known as the Krebs cycle, or citric acid cycle), and the electron transport chain, which uses cytochromes, flavins, and quinones as electron carriers. In some protozoans (and in nearly all other eukaryotes) the last two metabolic processes (the tricarboxylic acid cycle and electron transport) take place in mitochondria.

Aerobic protozoans are so small that they are able to obtain the oxygen they require for metabolism from the surrounding liquid medium by simple diffusion. The special pigments or structures required for the acquisition and transport of oxygen that are found in multicellular organisms are not required in protozoans. The pigment hemoglobin has been found in some ciliates (e.g., Tetrahymena), although it does not appear to function as an oxygen-carrying pigment as it does in humans. Within a single species, the rate of oxygen consumption varies in relation to factors such as temperature, the stage in the life cycle, and the cell's nutritional status (i.e., whether or not it is well fed).

Anaerobic Protozoans

Obligate anaerobes, in which metabolism must take place in the absence of oxygen, are rarely found among eukaryotic organisms. Those eukaryotes that are anaerobic often are either parasites or obligate symbionts of multicellular organisms that have evolved from aerobic ancestors. Excavata includes several anaerobic groups; many of its parasitic and symbiotic taxa live in the gastrointestinal tracts of invertebrates and vertebrates, such as humans. For instance, the diplomonad Giardia is an anaerobic parasite found in contaminated water that causes the gastrointestinal disease giardiasis. Trichomonads are a large group of anaerobic parasites. The organism Trichomonas vaginalis causes the sexually transmitted disease trichomoniasis. Trichomonads are closely related to the hypermastigotes, a group of anaerobes that are obligate symbionts of wood-digesting insects. Another large group of anaerobic symbionts of wood-digesting insects are the oxymonads. Some anaerobic protozoans are free-living.

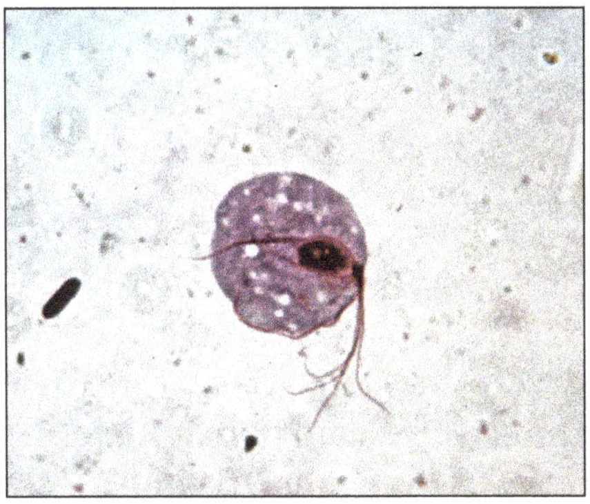

Trichomonas vaginalis A.L. Leu.

One ecological group of ciliates (e.g., Metopus, Plagiopyla, and Caenomorpha) is associated with sulfide-containing sediments. The sulfur ciliates harbour endosymbiotic and ectosymbiotic bacteria, which may take the metabolic end products released by the ciliates and reutilize them for growth and energy-yielding processes. Similar to other anaerobic protozoans, these ciliates are believed to have reverted from an aerobic metabolism to an anaerobic lifestyle in order to exploit a specialized ecological niche.

Hydrogenosomes

Unlike typical eukaryotic mitochondria, many anaerobic protozoans possess energy-yielding organelles belonging to a family of cellular structures called microbodies. Thus, these organisms do not perform the tricarboxylic acid cycle, nor do they possess electron transport chains. Instead, they must rely on substrate-level phosphorylation for the generation of the energy molecule adenosine triphosphate (ATP). A microbody commonly found in anaerobic protozoans is the hydrogenosome. Hydrogenosomes are enveloped by a double membrane and generate cellular energy via the partial oxidation of pyruvate to acetate (pyruvate fermentation). This reaction results in the production of carbon dioxide, molecular hydrogen (H_2), and ATP. The principle enzyme, pyruvate ferredoxin oxidoreductase, is present in high concentrations and forms conspicuous crystalline structures inside the hydrogenosome. The hydrogenosome is found in the trichomonads, hypermastigotes, and some euglenids. Hydrogenosomes are thought to have evolved from mitochondria.

Mitosomes and Glycosomes

Another type of anaerobic organelle common among anaerobic protozoans is the mitosome. The mitosome likely evolved from mitochondria, independent of hydrogenosome evolution. Mitosomes are found in diplomonads such as Giardia and were originally described in the intestinal parasite Entamoeba histolytica.

Certain parasitic protozoans that live in the blood, such as Trypanosoma brucei, have evolved a system of energy generation that makes use of yet another type of organelle, the glycosome. The glycosome contains glycolytic enzymes that oxidize glucose to the three-carbon molecule pyruvate. The glycosome is related to the peroxisome, a nearly ubiquitous eukaryotic organelle.

Carbon Acquisition and Nutrition

By definition, protozoans are nonfilamentous heterotrophs, meaning that they acquire carbon in the form of organic carbon from external sources, without the use of absorptive structures that are funguslike (i.e., hyphaelike). Instead, protozoans may ingest organic carbon substrates using phagotrophy. Food sources may include bacteria, algae, other protozoans, and small animals, such as the crustacean copepods.

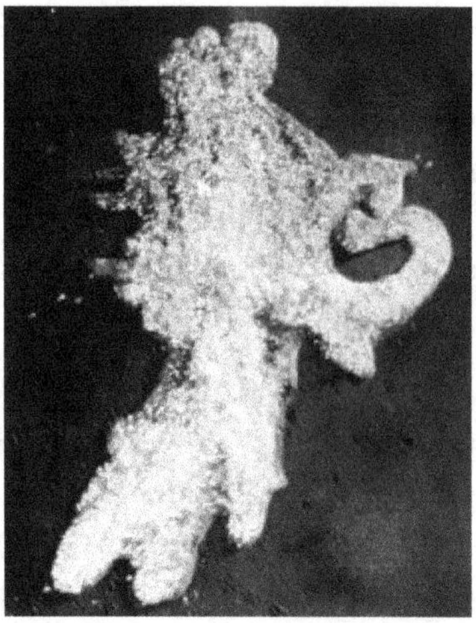

Amoeba engulfing a ciliate. Roman Vishniac Archive, International Center of Photography, New York, courtesy of Mara Vishniac Kohn.

Mechanisms of Food Ingestion

Protozoans may take food into the cell at a specific point, such as the cytostome (a well-developed feeding groove), at a particular region of the cell surface, or at any random point of entry. In the collared flagellates, or choanoflagellates, for example, the collar and flagellum operate in feeding. The collar, composed of fine pseudopodia, surrounds the flagellum. The beating flagellum creates a water current, causing water to move through the collar. Particles of food in the current are trapped on the collar and are ingested by pseudopodia at its base. The ingested food is then enclosed in a membrane-bound digestive or food vacuole.

Many ciliates are filter feeders, creating water currents with special ciliary structures associated with the cytostome. The synchronized beating of these ciliary structures pushes a stream of water against a membranelle composed of cilia. The membranelle acts as a collecting sieve, where the food particles become trapped in the free spaces between the cilia. Using this mode of feeding, ciliates can shift considerable volumes of water in relation to their size. Tetrahymena, for example, can filter 3,000 to 30,000 times its own volume in one hour.

Other ciliates lack complex oral cilia and gather their food by other means. Nassula has a complex cytostome and cytopharynx supported by a basketlike cytopharyngeal structure composed of microtubules. This species ingests filamentous algae by grasping the filament, bending it like a hairpin, and drawing it into the cytopharynx, where it is broken up into fragments and enclosed in digestive vacuoles. Predatory ciliates such as Didinium nasutum, Lacrymaria olor, and Dileptus anser apprehend their prey with special structures called extrusomes. Among the various types of extrusomes are the toxicysts, which are found in the oral region and release toxins that paralyze the prey.

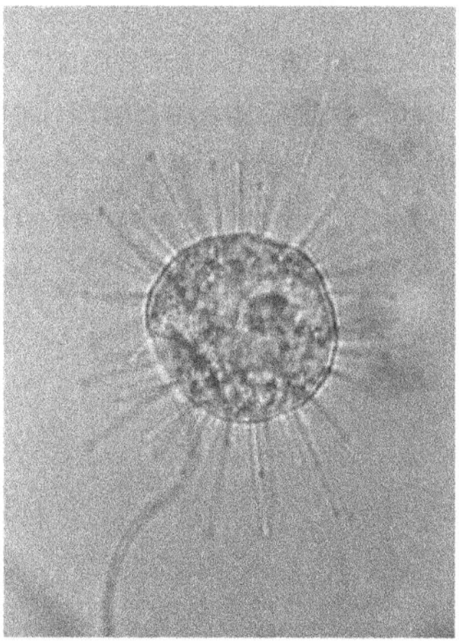

suctorian; Podophrya fixa The ciliated suctorian Podophrya fixa.

The suctorians are ciliate predators that usually possess tentacles of two functional types: feeding tentacles and piercing tentacles. The latter trap and immobilize the prey, usually other ciliates that make chance contact with the outstretched tentacles of the suctorian. The cell contents of the prey are transported up through the feeding tentacles into the suctorian, where digestive vacuoles are formed. The transporting mechanism is mediated by a complex array of microtubules within the tentacle. A single suctorian can often feed on several prey at the same time, and frequently the prey are larger than the predator.

Amoebae, all of which lack a cell mouth, or cytostome, also exhibit a diverse array of feeding mechanisms and diet. Some feed on filaments of cyanobacteria (blue-green algae)—which are composed of long chains of individual cells—by taking in the entire filament at any point on the cell surface and rolling it up into a coil inside a digestive vacuole. Others, such as the testate amoeba Pontigulasia, pierce single cells in algal filaments and remove the contents. The radiolarians and foraminiferans trap a wide range of prey, including protozoans, algae, and small crustaceans, in their complex pseudopodial networks and then transport the food items along the pseudopodia to the main body of the cell for ingestion. This process gives foraminiferans a distinctive appearance under the light microscope; relatively large grains of food particles are scattered across the reticulate pseudopodia, earning the organisms the name granuloreticulosans.

Parasitic protozoans feed in a variety of ways. Many live in the nutrient-rich medium of the body fluids—e.g., the blood or cells of their host. There they take in energy-rich fluids by pinocytosis, in which small amounts of the medium are pinched off into digestive vacuoles either at a specific site, such as the cytostome in ciliates or the flagellar pocket in trypanosomes, or along the surface of the cell in amoebae. Other parasitic protozoans engulf portions of the host tissue through phagocytosis in much the same way that free-living amoebae feed. Plasmodium, for example, engulfs portions of the red blood cells or liver cells in which they live. The hemoglobin in the cytoplasm of the red blood cell is only partially digested by the parasite. The protein portion of the hemoglobin molecule is degraded to its constituent amino acids, but the iron-containing portion is converted into insoluble iron-containing hemozoin, which remains within the parasite's endosomes until discarded at the next division. This process removes free hematin from the parasite cytoplasm, where it would otherwise prevent further metabolism within the parasite because it inhibits the actions of succinic dehydrogenase, an enzyme in the tricarboxylic acid cycle.

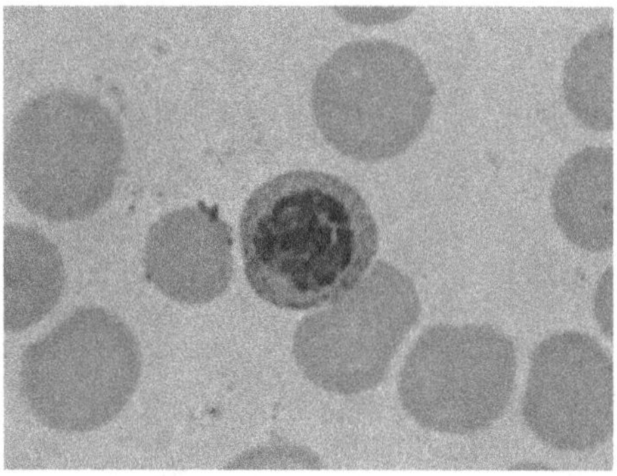

Plasmodium vivax in red blood cell.

Food Vacuoles

Whatever the mode of heterotrophic nutrition or diet, the food material is enclosed in food vacuoles, which are bounded by cell membrane. Digestive enzymes are poured into the newly formed vacuole from the surrounding cytoplasm. In the ciliate Paramecium, where the process has been researched in detail, the digestive vacuoles initially decrease in size and the enclosed particles aggregate. As digestion proceeds, the vacuole increases in size and the contents become progressively acidic, before gradually becoming alkaline near the end of the process. The products of digestion are then absorbed into the surrounding cytoplasm, and the waste material is ejected from the cell anus, or cytoproct. The length of the digestive cycle varies and depends on the species and the diet.

Paramecium contains a reservoir of membrane-forming material in discoid vesicles for the purpose of producing food vacuoles. The food vacuoles form at the cytopharynx when the cytopharyngeal membrane and the discoid vesicles fuse. At the cytoproct, where the vacuoles are broken down and the waste material of digestion is ejected, the membrane material is retrieved and returned to the cytopharynx. Thus, the pool of digestive vacuole membrane is continuously recycled within the cell.

Food Selection

While they seem to lack a sensory system, protozoans are capable of food selection. Many of the filter feeders apparently discriminate solely on the basis of size, dictated by the dimensions of the spaces in the membranelle acting as a sieve. Some filter-feeding ciliates, such as the tintinnids, however, are known to be selective and appear to be able to capture or reject items that arrive at the feeding membranelles in the feeding current. The large ciliate Stentor, for example, takes ciliates in preference to flagellated cells and algae, and discrimination increases as the animal becomes less hungry. Carnivorous species exercise distinct selectivity. Most suctorians feed exclusively on particular ciliate taxa. They are selective feeders and usually do not capture flagellates, amoebae, or their own ciliated swarmers. Evidence suggests that a reaction between chemical compounds on the surface of the prey and the tentacle tip of the suctorian is responsible for feeding selectivity. Amoeboid organisms also display feeding selectivity. Amoeba proteus, for example, selects the flagellate Chilomonas paramecium in preference to Monas punctum, even when the number of Monas in the medium is high. In this case, selection may be based on the digestibility of the prey; the latter is digested in 31/2 hours, the former in 3 to 18 minutes.

Mixotrophy

All protozoans engage in heterotrophy, but not all protozoans are exclusive heterotrophs. Those that combine autotrophy (self-sustaining food production from a carbon source and inorganic nitrogen) and heterotrophy (ingesting other organisms to acquire carbon) are known as mixotrophs. The degree of mixotrophy in a protozoan varies from complete reliance on the symbiotic alga (or algae) to transitory retention of the plastids of phytoflagellate prey with only a partial dependence on photosynthesis to supplement the cell's energy balance. For example, many protozoans, including the predatory ciliate Stentor and the heliozoan Acanthocystis, are capable of forming ephemeral symbioses with the coccoid green alga Chlorella. The ciliate Paramecium bursaria forms longer-lasting symbiosis with Chlorella but must nevertheless acquire the alga with each new generation.

Photosynthesis and Plastid Acquisition

Many coloured (i.e., photosynthetic) protists combine autotrophy with heterotrophy and therefore are mixotrophs. For example, some members of the euglenid and cryptomonad groups are mixotrophs. The mixotrophic members of these groups are commonly called acetate flagellates because their preferred organic carbon sources are acetates, simple fatty acids, and alcohols. These organisms are able to switch from carbohydrate-producing photosynthesis when light is available to heterotrophy on acetate and other substrates when light is not available.

In another form of mixotrophy, the amoebae and ciliates sequester only the plastids of their algal prey (rather than a complete algal cell) and use the plastids for photosynthesis. The plastids do not replicate inside the protozoan (as they do in the symbiotic algae), and thus they must be replaced continuously. The large marine ciliate Tontonia appendiculariformis may contain thousands of plastids that have been derived from a variety of flagellates. T. appendiculariformis, however, appears to be selective in its choice of prey, deriving plastids only from certain organisms.

Symbiotic Mixotrophy

Mixotrophy is a common phenomenon among free-living protozoans, which usually obtain the capability of photosynthesis from symbionts that are acquired with each new generation (i.e., the photosynthetic machinery is not inherited). Symbiotic mixotrophy, however, is not "true" mixotrophy, since the combined metabolism results from a consortium of two symbionts that are otherwise free-living.

Many of the foraminiferans and radiolarians possess symbiotic algae. In some foraminiferans and radiolarians several different symbiotic species of algae may live within the protozoan cytoplasm. During the day the endosymbionts are distributed in the pseudopodial network, but at night they are withdrawn close to the main body of the cell or into the shell. Many thousands of these algae may exist within a single protozoan, and a significant amount of the products of photosynthesis (e.g., glucose, alanine, maltose) are transferred from the algae to the protozoan. Indeed, in some circumstances, the protozoan can survive on this source of energy if deprived of food, although its growth may be impaired.

Mixotrophy in Planktonic Protozoans

Some mixotrophs (e.g., the planktonic protozoans Dinobryon and Ochromonas) also feed on bacteria but are phototrophs first and foremost. Thus, they are "true" mixotrophs, unlike protozoans that exploit the photosynthetic capability of symbiotic organisms and cannot carry out this form of metabolism themselves. Many planktonic marine and freshwater mixotrophs feed voraciously on bacteria. In some lakes these protists may be the main consumers of bacteria suspended in surface waters. It is believed that this ingestion of bacteria provides the mixotrophs not only with an additional source of carbon to supplement what is gained by photosynthesis but also with phosphorus and nitrogen, which are often scarce in biologically productive waters, and possibly with vitamins, which are essential to photosynthesis. Bacteria are more efficient at taking up these nutrients because they have a higher surface-to-volume ratio than protists. Thus, one way for the protists to acquire essential nutrients is to consume the bacteria.

Reproduction and Life Cycles

Asexual reproduction is the most common means of replication by protozoans. The ability to undergo a sexual phase is confined to the ciliates, the apicomplexans, and restricted taxa among the flagellated and amoeboid organisms. Moreover, sexual reproduction does not always result in an immediate increase in cell numbers but may simply be a means of exchanging genetic material between individuals of the same species (i.e., conjugation). Free-living protozoans normally resort to sexual reproduction only when environmental conditions become adverse, because this mode of reproduction enhances genetic variation through mechanisms such as mutation and chromosomal crossing over. These processes maintain genetic diversity within a population, which supports population fitness and survival. When food and other conditions are favourable, asexual reproduction occurs.

Mechanisms of Asexual Reproduction

Asexual reproduction in free-living species usually involves nuclear division and the division of the cell into two identical daughter cells of equal size by binary fission. In parasitic protozoans and some free-living species, multiple fission, resulting in the production of many offspring that may not resemble the parent cell, is normal. During the cycle of growth and division, the protozoan undergoes a series of identifiable phases: a division phase, a growth phase during which the cell increases substantially in size, a phase of DNA synthesis, and a phase of preparation for division, which extends from the end of DNA synthesis until the initiation of division. The division of the cytoplasm (cytokinesis) is preceded by the division of the nucleus or nuclei.

The plane of division in protozoan cells varies among the different groups and is of taxonomic significance. The ciliates normally divide in an equatorial, or transverse, plane, thereby maintaining the correct number of ciliary rows, or kineties. The cell mouth and any specialized cilia around it are replicated in different ways among the various ciliate groups, depending on the complexity of the cytostome. The replication of the cytostome precedes the division of the cytoplasm. Some ciliates (e.g., Colpoda) divide within thin-walled reproductive cysts into two daughter ciliates, each of which then divides so that the cyst contains four progeny, which are released when the cyst wall ruptures.

The sedentary suctorians do not reproduce by binary fission, because the production of an identical nonswimming offspring would rapidly lead to overcrowding. They instead produce single ciliated offspring, called swarmers, by a process called budding. Budding can occur endogenously, in which the bud forms within the parent and is ejected when mature, or exogenously, in which the swarmer is formed outside the parent. The swarmers swim away from the parent, settle on a substrate, lose their cilia, and develop feeding tentacles and an attaching stalk.

Naked amoebae have no fixed plane of division but simply round up and divide into two basically equal halves. The testate amoebae, which live in single-chambered shells, or tests, exude the daughter from the aperture of the shell. In species that have a shell formed from silica plates, the daughter contains the plates used to produce the shell but remains attached to the mother cell until the shell is fully formed, when the final severing of the cytoplasm between the individuals occurs. Some of the testate amoebae live inside proteinaceous shells. There too, the new shell is secreted before binary fission is completed.

The foraminiferan and radiolarian amoebae have evolved multiple fission. Both produce many flagellated swarmers, or zoospores. The common planktonic foraminiferan Globigerinoides sacculifer, for example, can produce 30,000 swarmers at one time. Each swarmer is about 5 micrometres (0.005 mm) long. In planktonic species the parent usually loses buoyancy and sinks by shedding spines and withdrawing the complicated pseudopodial network into the shell. The swarmers are produced in deep water and migrate upward as they mature. Each secretes a shell around itself, which is added to as the organism grows.

Mechanisms of Sexual Reproduction

The foraminiferans are unusual among free-living protozoans in that a sexual phase is a regular part of the life cycle, alternating with an asexual phase. During the life cycle two types of swarmers are produced. One type, zoospores, have half the number of chromosomes of the parent (i.e., they are haploid); they grow until they become mature adults and can produce and release large numbers of gametic swarmers. These gametes are identical (isogamous) but are functionally comparable to the eggs and sperm of higher organisms. The gametic swarmers fuse in pairs, thus restoring the full complement of chromosomes (i.e., they are diploid), and each individual grows, matures, and ultimately produces haploid zoospores.

Sexual reproduction among other protozoans is not widespread and can involve identical gametes (isogamy) or distinct male and female gametes (anisogamy or heterogamy). The female gametes are usually larger and are stationary, whereas the male gametes are smaller, produced in larger numbers, and motile.

Conjugation in Ciliates

Sexual reproduction among the ciliated protozoans takes the form of conjugation. The process does not result in an increase in numbers but is a simple exchange of genetic material between two individual cells. Conjugation occurs only between compatible mating strains within a species, and each species may contain many mating strains. Before conjugation occurs, special chemical signals, called gamones, are released by some ciliates. The gamones cause compatible mating strains to undergo processes that facilitate conjugation. In other ciliates, such as Paramecium, gamones are bound to the cell surface and elicit their responses when the ciliates make physical contact.

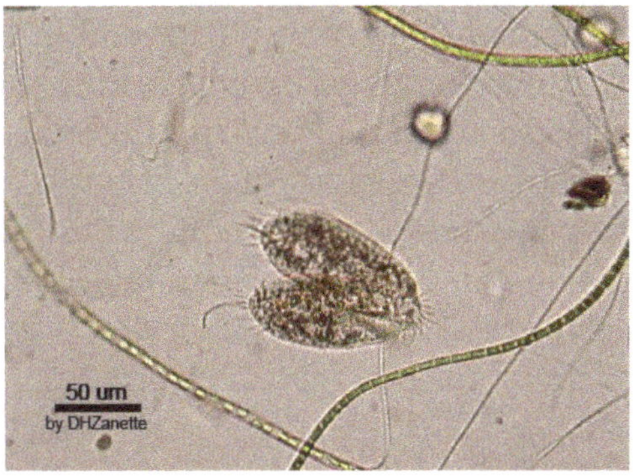

Conjugation Two hypotrichs in conjugation.

During conjugation, two ciliates line up side by side. The macronucleus, which plays no part in the process, disintegrates. A series of nuclear divisions of the micronuclei in each ciliate then ensues, including meiosis, during which a number of haploid micronuclei are produced in both cells. All but one of these haploid micronuclei disintegrate. The remaining haploid micronucleus in each cell then divides through mitosis into two haploid nuclei (gamete nuclei). A bridge of cytoplasm forms between the two ciliates, and one gametic nucleus from each cell passes into the other cell. The two gametic nuclei in each cell unite, thus restoring the diploid number of chromosomes. The micronucleus undergoes two mitotic divisions to produce four micronuclei: two of these will form the new micronuclei of the cell, and two are destined to become the macronucleus. Following the process of conjugation, normal binary fission proceeds. The number of macronuclei and micronuclei formed is dependent on the species and remains the same as the original number.

Autogamy and Modified Conjugation

When no suitable mating partner is available, ciliates may undergo a form of conjugation called autogamy, in which all the nuclear processes described above occur. But, because only one individual is involved, there is no exchange of gametic nuclei. Instead, the two gametic nuclei within the cell unite to form the restored micronucleus.

Specialized sedentary suctorian ciliates practice a modified form of conjugation. The conjugating individuals differ in appearance. The macroconjugants resemble the normal feeding individuals, and the microconjugants resemble the swarmers, although smaller. When a microconjugant locates a macroconjugant, it enters and fuses with it. This is quite different from the temporary association between two cells that occurs during sexual reproduction in most ciliates.

Parasitic Protozoan Life Cycles

As is common with other parasitic organisms, parasitic protozoans face the problem of how to disperse from one host to another. In order to increase the probability of finding more hosts, most parasitic protozoans reproduce in high numbers. A representative life cycle of a parasitic protozoan can be found in members of the parasitic group Apicomplexa. These protozoans have a complex life cycle that involves a series of stages characterized by episodes of asexual multiple division called schizogony. In the parasite Plasmodium, for example, this phase of the life cycle occurs in the liver and red blood cells of humans. The parasite (sporozoite) enters the host's cells and grows while feeding on the cell contents. It then undergoes a multiple asexual division (schizogony) into many individuals (merozoites). The host's cell wall ruptures, permitting each individual to invade a new red blood cell and repeat the process.

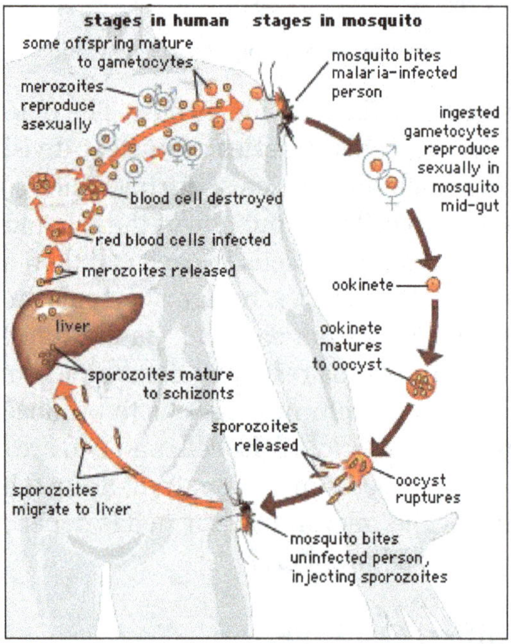

Life cycle of a malaria parasite.

In certain merozoites a sexual cycle is eventually initiated inside the red blood cell, and male and female gametes are produced. The male gametes (microgametocytes) are small, while the female gametes (macrogametocytes) are larger. The life cycle continues if the gametocytes are taken up by a feeding female mosquito of the genus Anopheles. Only the gametocytes can infect the mosquito. Inside the mosquito's gut the haploid gametes fuse to form a diploid zygote, which then undergoes sporogony, a process of multiple divisions in which many sporozoites are produced. The sporozoites migrate to the salivary glands of the insect and are injected into a new host when the mosquito feeds again. They are carried by the blood to the liver, where they undergo their first schizogony inside liver cells, thereafter invading the red blood cells for repeated cycles of schizogony.

Flagellated protozoan parasites reproduce almost exclusively by asexual means and do not appear to have a sexual phase in their life cycles. There is, however, evidence of genetic exchange between certain subspecies of Trypanosoma brucei.

Adaptations

For the most part, parasitic protozoans live in a fairly constant environment. Temperature fluctuates very little, or not at all, inside the host, desiccation is not a risk, and food is in constant supply. Free-living protists, on the other hand, face short- or long-term changes in temperature, aquatic acidity, food supply, moisture, and light. Many protozoans respond to adverse environmental conditions by encysting: they secrete a thick, tough wall around themselves and effectively enter a quiescent state comparable to hibernation. The ability to form a resistant cyst is widespread among diverse protistan groups and probably developed early in their evolutionary history. Resting cysts also are easily carried by the wind and form an important means of dispersal for species that live in the soil or are common in ephemeral ponds and pools. In climates with distinct cold seasons, the cyst may be an important phase in the annual life cycle.

The cyst wall is composed of a varying number of layers, the components of which are dependent on the species. During the encystment process, the protozoan cell undergoes a series of changes that considerably reduce the complexity of the organism. Flagellated organisms and ciliates lose their flagella and cilia, the contractile vacuole and food vacuoles disappear, and the distribution of organelles within the cell may be reorganized. In some species the cell volume reduces considerably. These changes are reversed during the process of excystment.

Certain marine planktonic tintinnids are programmed to break out of their cysts en masse at times of the year when the food supply is abundant. Helicostomella subulata, for example, excysts in June in temperate waters and becomes numerous from July through October. It encysts again in October, sinking to the sediments, where it remains until the following year. The cyst is a normal part of the annual life cycle, and even laboratory populations of this ciliate encyst at the same time as the natural population. This type of life strategy pattern has been demonstrated in several other ciliates and in some amoebae.

For soil-dwelling protozoans the cyst is an important refuge when soil moisture disappears or when soil water becomes frozen. In soils that are subject to freezing and periodic short-term thawing, the protozoans rapidly excyst, feed, and reproduce and then encyst again when soil water becomes temporarily unavailable to them.

The cyst plays an important role in the life cycles of several parasitic protozoans that have a free-living dispersal stage, such as Entamoeba histolytica and Cryptosporidium.

The cysts are excreted in the host's feces and survive in water or soil. Humans are usually infected through drinking contaminated water or eating raw fruit and vegetables grown where human feces are used as fertilizer.

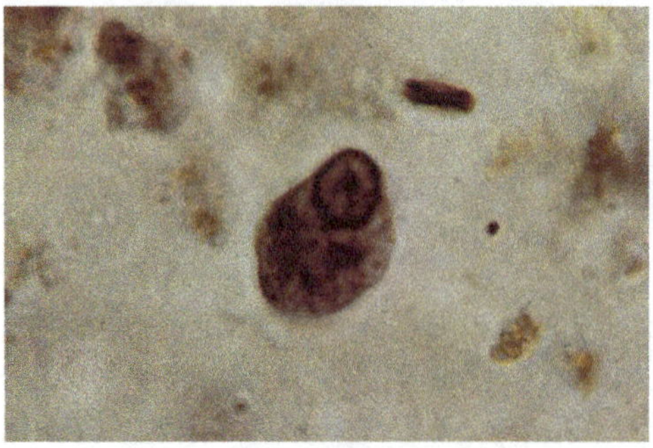

Entamoeba coli.

Some freshwater protozoans, especially the ciliates Spirostomum, Loxodes, and Plagiopyla, avoid unpleasant conditions, especially lack of oxygen, by abandoning their bottom-dwelling way of life and swimming upward to position themselves at a level where some oxygen is available but where they are not in direct competition with planktonic species. They remain there until oxygen again becomes available on the lake bottom, at which time they migrate downward.

The widespread occurrence of mixotrophy involving algal symbiosis and the retention and sequestration of the plastids of photosynthetic prey by planktonic protozoans is believed to be an adaptation to waters where food is limited. Ciliates that retain plastids appear to be far more common in waters where food is scarce than in productive waters. An inverse relationship exists between this form of mixotrophy and the productivity of the ecosystem.

Classification

General Principles

A fundamental shift in protozoan taxonomy occurred in 1990, when American microbiologist Carl Woese and colleagues revolutionized the world of biology with the three-domain classification system of life. Based on sequences of rRNA (ribosomal RNA), molecules present in all organisms as part of the protein-manufacturing machinery, Woese's classification system revealed three major evolutionary groups of life on Earth, one of which is eukaryotic (the Eukarya) and two of which are prokaryotic (Eubacteria and Archaea). It is generally accepted that the Eubacteria (now Bacteria) are the most distant genetic group of the three. The three-domain system has largely replaced the previous five-kingdom system of American biologist Robert H. Whittaker, which is based on morphology (e.g., mode of nutrition) rather than phylogeny (the history of the evolution of a species or group).

Classification within the domain Eukarya also experienced a shift from morphology-based approaches to an emphasis on phylogenetic relationships. On the basis of that advance, the former eukaryotic kingdoms Animalia, Plantae, and Fungi are no longer separated from protists. Instead, each of those major groups of multicellular life is classified within a supergroup of Eukarya along with a protistan group. For example, plants are classified within the supergroup Archaeplastida, alongside some examples of unicellular algae, and the animals and fungi are classified within the supergroup Opisthokonta, along with single-celled choanoflagellates.

Current scientific approaches have produced a nested yet nonhierarchical picture of biological classification, in stark contrast to the heirarchical scheme of Linnean taxonomy, which specifies somewhat arbitrary universal ranks of classification (e.g., Kingdom, Phylum, Class, Order). This development has caused many biologists to abandon the Linnean system, primarily at the higher levels of classification, rather than at the genus and species levels. The validity and utility of the Linnean taxonomic approach continues to be a source of debate among biologists. Major developments in the classification of protists in the 1990s and early 2000s did not use hierarchical schemes.

Protistan systematics remains a subject of debate and change. Protists comprise a large and sometimes unwieldy assemblage, and assignments of species to particular taxa change as new genomic approaches, biochemical techniques, and electron microscopy studies provide more details on the affinities of various species.

Diagnostic Features

A general consensus regarding the classification of eukaryotes (with emphasis on protists) was published by the International Society of Protistologists (ISOP) in 2005. This classification system of eukaryotic taxa divides species into monophyletic groups, or clades. Monophyletic groups contain a common ancestor and all its descendants. This type of grouping is in contrast to paraphyletic groupings (consisting of a common ancestor and some of its descendants) and polyphyletic groupings (consisting of taxa that do not share a common ancestor). Monophyletic groups are defined by the possession of shared, derived traits known as apomorphies.

The classification scheme introduced by the ISOP defines six monophyletic supergroups of eukaryotes: Archaeplastida, Excavata, Chromalveolata, Amoebozoa, Rhizaria, and Opisthokonta. Using this scheme, the protozoans and algae are clearly polyphyletic. Former groupings defined by previous classification schemes are no longer recognized in the 2005 system, such as the former phylum Sarcomastigophora, which grouped paraphyletic lineages of photosynthetic and nonphotosynthetic taxa.

VIRUS

Virus is an infectious agent of small size and simple composition that can multiply only in living cells of animals, plants, or bacteria.

The earliest indications of the biological nature of viruses came from studies in 1892 by the Russian scientist Dmitry I. Ivanovsky and in 1898 by the Dutch scientist Martinus W. Beijerinck. Beijerinck first surmised that the virus under study was a new kind of infectious agent, which he designated contagium vivum fluidum, meaning that it was a live, reproducing organism that differed from other organisms. Both of these investigators found that a disease of tobacco plants could be transmitted by an agent, later called tobacco mosaic virus, passing through a minute filter that would not allow the passage of bacteria. This virus and those subsequently isolated would not grow on an artificial medium and were not visible under the light microscope. In independent studies in 1915 by the British investigator Frederick W. Twort and in 1917 by the French Canadian scientist Félix H. d'Hérelle, lesions in cultures of bacteria were discovered and attributed to an agent called bacteriophage ("eater of bacteria"), now known to be viruses that specifically infect bacteria.

The unique nature of these organisms meant that new methods and alternative models had to be developed to study and classify them. The study of viruses confined exclusively or largely to humans, however, posed the formidable problem of finding a susceptible animal host. In 1933 the British investigators Wilson Smith, Christopher H. Andrewes, and Patrick P. Laidlaw were able to transmit influenza to ferrets, and the influenza virus was subsequently adapted to mice. In 1941 the American scientist George K. Hirst found that influenza virus grown in tissues of the chicken embryo could be detected by its capacity to agglutinate (draw together) red blood cells.

A significant advance was made by the American scientists John Enders, Thomas Weller, and Frederick Robbins, who in 1949 developed the technique of culturing cells on glass surfaces; cells could then be infected with the viruses that cause polio (poliovirus) and other diseases. (Until this time, the poliovirus could be grown only in the brains of chimpanzees or the spinal cords of monkeys.) Culturing cells on glass surfaces opened the way for diseases caused by viruses to be identified by their effects on cells (cytopathogenic effect) and by the presence of antibodies to them in the blood. Cell culture then led to the development and production of vaccines (preparations used to elicit immunity against a disease) such as the poliovirus vaccine.

Scientists were soon able to detect the number of bacterial viruses in a culture vessel by measuring their ability to break apart (lyse) adjoining bacteria in an area of bacteria (lawn) overlaid with an inert gelatinous substance called agar—viral action that resulted in a clearing, or "plaque." The American scientist Renato Dulbecco in 1952 applied this technique to measuring the number of animal viruses that could produce plaques in layers of adjoining animal cells overlaid with agar. In the 1940s the development of the electron microscope permitted individual virus particles to be seen for the first time, leading to the classification of viruses and giving insight into their structure.

Advancements that have been made in chemistry, physics, and molecular biology since the 1960s have revolutionized the study of viruses. For example, electrophoresis on gel substrates gave a deeper understanding of the protein and nucleic acid composition of viruses. More-sophisticated immunologic procedures, including the use of monoclonal antibodies directed to specific antigenic sites on proteins, gave a better insight into the structure and function of viral proteins. The progress made in the physics of crystals that could be studied by X-ray diffraction provided the high resolution required to discover the basic structure of minute viruses. Applications of new knowledge about cell biology and biochemistry helped to determine how viruses use their host cells for synthesizing viral nucleic acids and proteins.

The revolution that took place in the field of molecular biology allowed the genetic information encoded in nucleic acids of viruses—which enables viruses to reproduce, synthesize unique proteins, and alter cellular functions—to be studied. In fact, the chemical and physical simplicity of viruses has made them an incisive experimental tool for probing the molecular events involved in certain life processes. Their potential ecological significance was realized in the early 21st century, following the discovery of giant viruses in aquatic environments in different parts of the world.

This article discusses the fundamental nature of viruses: what they are, how they cause infection, and how they may ultimately cause disease or bring about the death of their host cells.

General Features

Viruses occupy a special taxonomic position: they are not plants, animals, or prokaryotic bacteria (single-cell organisms without defined nuclei), and they are generally placed in their own kingdom. In fact, viruses should not even be considered organisms, in the strictest sense, because they are not free-living; i.e., they cannot reproduce and carry on metabolic processes without a host cell.

All true viruses contain nucleic acid—either DNA (deoxyribonucleic acid) or RNA (ribonucleic acid)—and protein. The nucleic acid encodes the genetic information unique for each virus. The infective, extracellular (outside the cell) form of a virus is called the virion. It contains at least one unique protein synthesized by specific genes in the nucleic acid of that virus. In virtually all viruses, at least one of these proteins forms a shell (called a capsid) around the nucleic acid. Certain viruses also have other proteins internal to the capsid; some of these proteins act as enzymes, often during the synthesis of viral nucleic acids. Viroids (meaning "viruslike") are disease-causing organisms that contain only nucleic acid and have no structural proteins. Other viruslike particles called prions are composed primarily of a protein tightly complexed with a small nucleic acid molecule. Prions are very resistant to inactivation and appear to cause degenerative brain disease in mammals, including humans.

Viruses are quintessential parasites; they depend on the host cell for almost all of their life-sustaining functions. Unlike true organisms, viruses cannot synthesize proteins, because they lack ribosomes (cell organelles) for the translation of viral messenger RNA (mRNA; a complementary copy of the nucleic acid of the nucleus that associates with ribosomes and directs protein synthesis) into proteins. Viruses must use the ribosomes of their host cells to translate viral mRNA into viral proteins.

Viruses are also energy parasites; unlike cells, they cannot generate or store energy in the form of adenosine triphosphate (ATP). The virus derives energy, as well as all other metabolic functions, from the host cell. The invading virus uses the nucleotides and amino acids of the host cell to synthesize its nucleic acids and proteins, respectively. Some viruses use the lipids and sugar chains of the host cell to form their membranes and glycoproteins (proteins linked to short polymers consisting of several sugars).

The true infectious part of any virus is its nucleic acid, either DNA or RNA but never both. In many viruses, but not all, the nucleic acid alone, stripped of its capsid, can infect (transfect) cells, although considerably less efficiently than can the intact virions.

The virion capsid has three functions: (1) to protect the viral nucleic acid from digestion by certain enzymes (nucleases), (2) to furnish sites on its surface that recognize and attach (adsorb) the virion to receptors on the surface of the host cell, and, in some viruses, (3) to provide proteins that form part of a specialized component that enables the virion to penetrate through the cell surface membrane or, in special cases, to inject the infectious nucleic acid into the interior of the host cell.

Host Range and Distribution

Logic originally dictated that viruses be identified on the basis of the host they infect. This is justified in many cases but not in others, and the host range and distribution of viruses are only one criterion for their classification. It is still traditional to divide viruses into three categories: those that infect animals, plants, or bacteria.

Virtually all plant viruses are transmitted by insects or other organisms (vectors) that feed on plants. The hosts of animal viruses vary from protozoans (single-celled animal organisms) to humans. Many viruses infect either invertebrate animals or vertebrates, and some infect both. Certain viruses that cause serious diseases of animals and humans are carried by arthropods. These vector-borne viruses multiply in both the invertebrate vector and the vertebrate host.

Certain viruses are limited in their host range to the various orders of vertebrates. Some viruses appear to be adapted for growth only in ectothermic vertebrates (animals commonly referred to as cold-blooded, such as fishes and reptiles), possibly because they can reproduce only at low temperatures. Other viruses are limited in their host range to endothermic vertebrates (animals commonly referred to as warm-blooded, such as mammals).

Size and Shape

The amount and arrangement of the proteins and nucleic acid of viruses determine their size and shape. The nucleic acid and proteins of each class of viruses assemble themselves into a structure called a nucleoprotein, or nucleocapsid. Some viruses have more than one layer of protein surrounding the nucleic acid; still others have a lipoprotein membrane (called an envelope), derived from the membrane of the host cell, that surrounds the nucleocapsid core. Penetrating the membrane are additional proteins that determine the specificity of the virus to host cells. The protein and nucleic acid constituents have properties unique for each class of virus; when assembled, they determine the size and shape of the virus for that specific class. The genomes of Mimiviruses and Pandoraviruses, which are some of the largest known viruses, range from 1 to 2.5 Mb (1 Mb = 1,000,000 base pairs of DNA).

Most viruses vary in diameter from 20 nanometres (nm; 0.0000008 inch) to 250–400 nm; the largest, however, measure about 500 nm in diameter and are about 700–1,000 nm in length. Only the largest and most complex viruses can be seen under the light microscope at the highest resolution. Any determination of the size of a virus also must take into account its shape, since different classes of viruses have distinctive shapes.

Shapes of viruses are predominantly of two kinds: rods, or filaments, so called because of the linear array of the nucleic acid and the protein subunits; and spheres, which are actually 20-sided (icosahedral) polygons. Most plant viruses are small and are either filaments or polygons, as are many bacterial viruses. The larger and more-complex bacteriophages, however, contain as their genetic information double-stranded DNA and combine both filamentous and polygonal shapes. The classic T4 bacteriophage is composed of a polygonal head, which contains the DNA genome and a special-function rod-shaped tail of long fibres. Structures such as these are unique to the bacteriophages.

Animal viruses exhibit extreme variation in size and shape. The smallest animal viruses belong to the families Parvoviridae and Picornaviridae and measure about 20 nm and about 30 nm in diameter, respectively. Viruses of these two families are icosahedrons and contain nucleic acids with limited genetic information. Viruses of the family Poxviridae are about 250 to 400 nm in their longest dimension, and they are neither polygons nor filaments. Poxviruses are structurally more complex than simple bacteria, despite their close resemblance. Animal viruses that have rod-shaped (helical) nucleocapsids are those enclosed in an envelope; these viruses are found in the families Paramyxoviridae, Orthomyxoviridae, Coronaviridae, and Rhabdoviridae. Not all enveloped viruses contain helical nucleocapsids, however; those of the families Herpesviridae, Retroviridae, and Togaviridae have polygonal nucleocapsids. Most enveloped viruses appear to be spherical, although the rhabdoviruses are elongated cylinders.

The criteria used for classifying viruses into families and genera are primarily based on three structural considerations: (1) the type and size of their nucleic acid, (2) the shape and size of the capsids, and (3) the presence of a lipid envelope, derived from the host cell, surrounding the viral nucleocapsid.

The Nucleic Acid

As is true in all forms of life, the nucleic acid of each virus encodes the genetic information for the synthesis of all proteins. In almost all free-living organisms, the genetic information is in the form of double-stranded DNA arranged as a spiral lattice joined at the bases along the length of the molecule (a double helix). In viruses, however, genetic information can come in a variety of forms, including single-stranded or double-stranded DNA or RNA.

The nucleic acids of virions are arranged into genomes. All double-stranded DNA viruses consist of a single large molecule, whereas most double-stranded RNA viruses have segmented genomes, with each segment usually representing a single gene that encodes the information for synthesizing a single protein. Viruses with single-stranded genomic DNA are usually small, with limited genetic information. Some single-stranded DNA viruses are composed of two populations of virions, each consisting of complementary single-stranded DNA of polarity opposite to that of the other.

The virions of most plant viruses and many animal and bacterial viruses are composed of single-stranded RNA. In most of these viruses, the genomic RNA is termed a positive strand because the genomic RNA acts as mRNA for direct synthesis (translation) of viral protein. Several large families of animal viruses, and one that includes both plant and animal viruses (the Rhabdoviridae), however, contain genomic single-stranded RNA, termed a negative strand, which is complementary to mRNA. All of these negative-strand RNA viruses have an enzyme, called an RNA-dependent RNA polymerase (transcriptase), which must first catalyze the synthesis of complementary mRNA from the virion genomic RNA before viral protein synthesis can occur. These variations in the nucleic acids of viruses form one central criterion for classification of all viruses.

A distinctive large family of single-stranded RNA viruses is called Retroviridae; the RNA of these viruses is positive, but the viruses are equipped with an enzyme, called a reverse transcriptase, that copies the single-stranded RNA to form double-stranded DNA.

The Protein Capsid

The protein capsid provides the two free ends is essentially completely extended or somewhat coiled (a helix) and (2) those in which the nucleic acid, which may or may not be a covalently closed circle, is wound tightly into a condensed configuration, like a ball of yarn. These two classes of virus assume in the first case a long, extended rodlike structure and in the second case a symmetrical polygon.

By far the best-studied example of a helical rod-shaped virus is the tobacco mosaic virus, which was crystallized by Wendell Stanley in 1935. The tobacco mosaic virus contains a genome of single-stranded RNA encased by 2,130 molecules of a single protein; there are 161/3 protein molecules for each turn of the RNA helix in the ratio of three nucleotides for each protein molecule.

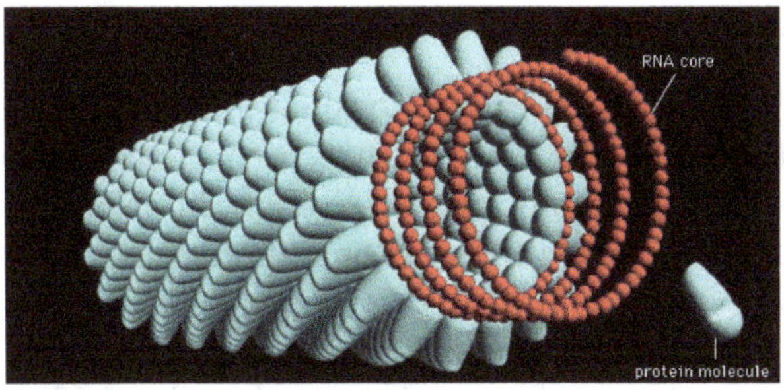

Schematic structure of the tobacco mosaic virus. The cutaway section shows the helical ribonucleic acid associated with protein molecules in a ratio of three nucleotides per protein molecule.

Under the right environmental conditions, viral RNA and protein molecules in liquid suspension will assemble themselves into a perfectly formed and fully infectious virus. The length of the helical virus capsid is determined by the length of the nucleic acid molecule, which is the framework for the assembly of the capsid protein. The various helical viruses have different lengths and widths, depending on the size of the nucleic acid as well as on the mass and shape of the protein molecule. Some of these helical viruses form rigid rods, while others form flexible rods, depending on the properties of the assembled proteins.

Polygonal viruses vary greatly in size, from 20 to 150 nm in diameter, essentially proportional to the size of the nucleic acid molecule coiled up inside the virion. Most, if not all, of the polygonal viruses are icosahedral; like a geodesic dome, they are formed by equilateral triangles, in this case 20. Each triangle is composed of protein subunits (capsomeres), often in the form of hexons (multiples of six) that are the building blocks of the capsid. There are 12 vertices (the apical junctions of these 20 triangles), each comprising a penton (five subunits). These icosahedral virions have three axes of fivefold, threefold, and twofold rotational symmetry. The number of capsomeres is a basis for taxonomic classification of these virus families. Certain icosahedral viruses, usually those that are more complex, contain internal proteins adhering to the nucleic acid that are not accessible at the surface of the virions.

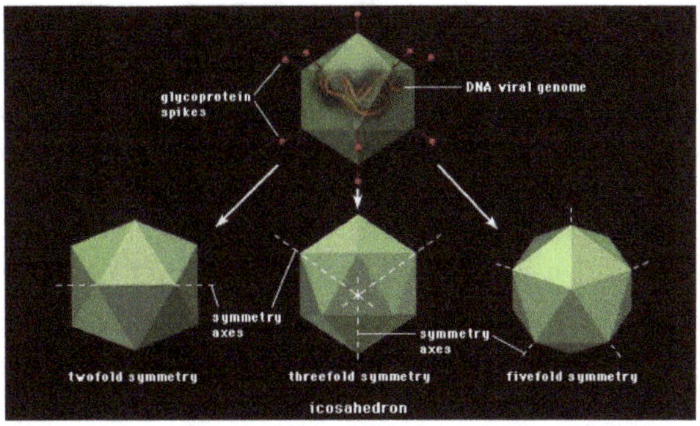

A virus icosahedron (20-sided structure) shown in the (left) twofold, (centre) threefold, and (right) fivefold axes of symmetry. Edges of the upper and lower surfaces are drawn in solid and broken lines, respectively.

The Lipoprotein Envelope

Surrounding viruses of either helical or icosahedral symmetry are lipoprotein envelopes, unit membranes of two lipid layers interspersed with protein molecules (lipoprotein bilayer). These viral membranes are composed of phospholipids and neutral lipids (largely cholesterol) derived from cell membranes during the process known as budding. Virtually all proteins of the cell membrane, however, are replaced by proteins of viral origin during budding. Although all the viral envelope lipids originate from the cell, their relative proportions vary from those in the cell membrane because the viral proteins select only certain lipids during budding.

Associated with the virion membrane are "integral" glycoproteins, which completely traverse the lipid bilayer, and "peripheral" matrix proteins, which line the inner surface. The glycoproteins contain regions of amino acids that, in the first step of viral infection, recognize host-cell receptors. Matrix proteins appear to function in the selection of regions of the cell membrane to be used for the viral membrane, as well as to aid the virion in entering cells.

The Cycle of Infection

Viruses can reproduce only within a host cell. The parental virus (virion) gives rise to numerous progeny, usually genetically and structurally identical to the parent virus. The actions of the virus depend both on its destructive tendencies toward a specific host cell and on environmental conditions. In the vegetative cycle of viral infection, multiplication of progeny viruses can be rapid. This cycle of infection often results in the death of the cell and the release of many virus progeny. Certain viruses, particularly bacteriophages, are called temperate (or latent) because the infection does not immediately result in cell death. The viral genetic material remains dormant or is actually integrated into the genome of the host cell. Cells infected with temperate viruses are called lysogenic because the cells tend to be broken down when they encounter some chemical or physical factor, such as ultraviolet light. In addition, many animal and plant viruses, the genetic information of which is not integrated into the host DNA, may lie dormant in tissues for long periods of time without causing much, if any, tissue damage. Viral infection does not always result in cell death or tissue injury; in fact, most viruses lie dormant in tissue without ever causing pathological effects, or they do so only under other, often environmental, provocations.

Although the reproductive pathways of different viruses vary considerably, there are certain basic principles and a particular series of events in the cycle of infection for most, if not all, viruses. The first step in the cycle of infection is that the invading parental virus (virion) must attach to the surface of the host cell (adsorption). In the second step, the intact virion either penetrates the outer membrane and enters the cell's interior (cytoplasm) or injects the genetic material of the virus into the interior of the cell while the protein capsid (and envelope, if present) remains at the cell surface. In the case of whole-virion penetration, a subsequent process (uncoating) liberates the genetic material from the capsid and envelope, if present. In either case, the viral genetic material cannot begin to synthesize protein until it has emerged from the capsid or envelope.

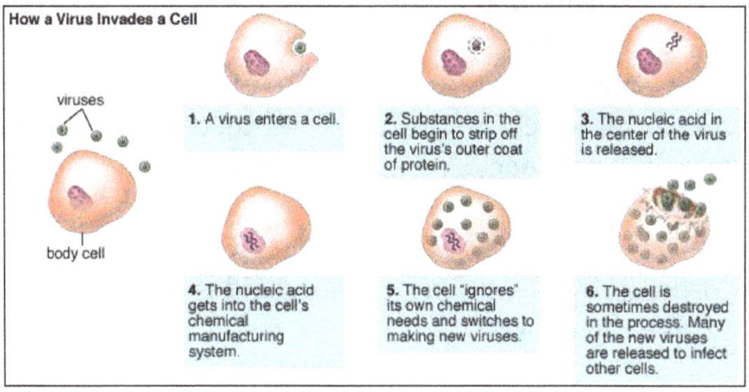

virus: invasion of a cell The process by which a virus invades a cell and reproduces.

Certain bacterial viruses, such as the T4 bacteriophage, have evolved an elaborate process of infection: following adsorption and firm attachment of the virus's tail to the bacterium surface by means of proteinaceous "pins," the musclelike tail contracts, and the tail plug penetrates the cell wall and underlying membrane and injects virus (phage) DNA into the cell. Other bacteriophages penetrate the cell membrane by different means, such as injecting the nucleic acid through the male (sex) pili of the bacterium. In all bacterial viruses, penetration transmits the viral nucleic acid through a rigid bacterial cell wall.

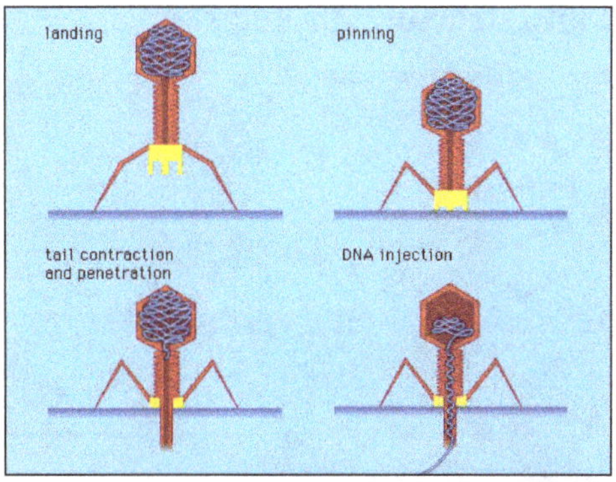

General structure of T4 bacteriophage and a model of its mode of attachment to, and injection of its DNA into, a bacterial cell.

Plant cells also have rigid cell walls, which plant viruses cannot ordinarily penetrate. Plant viruses, however, have not evolved their own systems for injecting nucleic acids into host cells, and so they are transmitted by the proboscis of insects that feed on plants. In the laboratory, plant viruses penetrate plant cells if the cell walls have been abraded with sandpaper or if cell protoplasts (plasma membrane, cytoplasm, and nucleus) are devoid of walls.

Penetration of animal cells by viruses involves different processes, because animal cells are enclosed not by walls but by a flexible lipoprotein bilayer membrane. Most animal viruses, whether or not they are encased in lipid envelopes, penetrate cells in an intact form by a process called endocytosis. The membrane invaginates and engulfs a virus particle adsorbed to a cell, usually in an area of the membrane called a coated pit, which is lined by a special protein known as clathrin. As the coated pit invaginates, it is pinched off in the cytoplasm to form a coated vesicle. The coated vesicle fuses with cytoplasmic endosomes (membrane-enclosed vesicles) and then with cell organelles called lysosomes, which are membrane-enclosed vesicles containing enzymes. In an acidic environment, the membrane of an enveloped virus fuses with the endosome membrane, and the viral nucleocapsid is released into the cytoplasm. Nonenveloped viruses presumably undergo a similar process, by which the protein capsid is degraded, releasing the naked viral nucleic acid into the cytoplasm.

The order of the stages of viral replication that follow the uncoating of the genome varies for different virus classes. For many virus families the third step in the cycle of infection is transcription of the genome of the virus to produce viral mRNA, followed by the fourth step, translation of viral mRNA into proteins. For those viruses in which the genomic nucleic acid is an RNA that can serve as a messenger (i.e., positive-strand RNA viruses), the third step is the translation of the RNA to form viral proteins; some of these newly synthesized viral proteins are enzymes that synthesize nucleic acids (polymerases), which carry out a fourth step, the transcription of more mRNA from the viral genome. For the more complicated DNA viruses, such as adenoviruses and herpesviruses, some regions of the genome synthesize "early" mRNAs, which are translated into polymerases that initiate the transcription of "late" regions of the DNA into mRNAs, which are then translated into structural proteins.

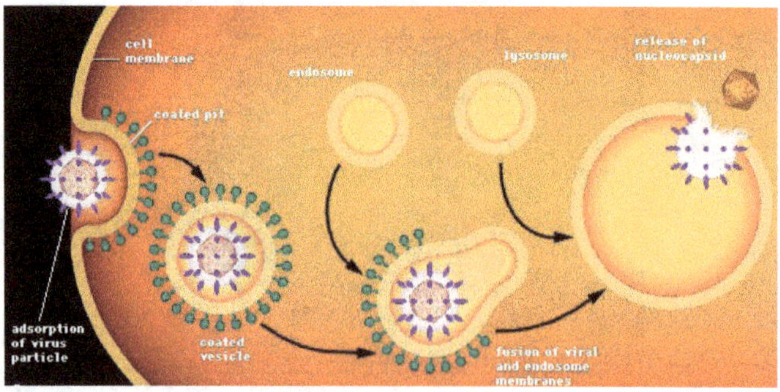

Adsorption to and entry into a cell of an enveloped animal virus by the process of endocytosis into clathrin-coated vesicles, which fuse with large vesicles (endosomes and lysosomes).

Regardless of how the third and fourth steps proceed, the fifth step in the cycle of infection is replication (reproduction of the parental genome to make progeny genomes). The sixth step is the assembly of the newly replicated progeny genomes with structural proteins to make fully formed progeny virions. The seventh and last step is the release of progeny virions by lysis of the host cell through the process of either extrusion or budding, depending on the nature of the virus. In a host animal or cell culture, this seven-step process may be repeated many times; the progeny virions released from the original site of infection are then transmitted to other sites or to other individuals.

For most animal and plant RNA viruses, all replicative events take place in the cytoplasm; in fact, many of these RNA viruses can grow in host cells in which the nucleus has been removed. Replication of most animal and plant DNA viruses, as well as the RNA influenza virus, takes place in the nucleus. In these viruses, transcription takes place in the nucleus, the mRNA migrates to the cytoplasm, where it is translated, and these viral proteins migrate back to the nucleus, where they assemble with newly replicated progeny genomes. Migration of newly translated viral proteins from the cytoplasm to the nucleus is generally a function of specific amino acid sequences called "signals," which translocate the protein through pores in the nucleus membrane.

Viral DNA Integration

Lysogeny

Many bacterial and animal viruses lie dormant in the infected cell, and their DNA may be integrated into the DNA of the host cell chromosome. The integrated viral DNA replicates as the cell genome replicates; after cell division, the integrated viral DNA is duplicated and usually distributed equally to the two cells that result. The bacteria that carry the noninfective precursor phage, called the prophage, remain healthy and continue to grow until they are stimulated by some perturbing factor, such as ultraviolet light. The prophage DNA is then excised from the bacterial chromosome, and the phage replicates, producing many progeny phages and lysing the host bacterial cell. This process, originally discovered in temperate bacteriophages in 1950 by the French microbiologist André Lwoff, is called lysogeny.

The classic example of a temperate bacteriophage is called lambda (λ) virus, which readily causes lysogeny in certain species of the bacterium Escherichia coli. The DNA of the λ bacteriophage is integrated into the DNA of the E. coli host chromosome at specific regions called attachment sites. The integrated prophage is the inherited, noninfectious form of the virus; it contains a gene that represses the lytic functions of the phage and thus ensures that the host cell will continue to replicate the phage DNA along with its own and that it will not be destroyed by the virus. Ultraviolet light, or other factors that stimulate the replication of DNA in the host cell, causes the formation of a recA protease, an enzyme that breaks apart the λ phage repressor and induces λ phage replication and, eventually, destruction of the host cell.

Excision of the prophage DNA from the host chromosomal DNA (as an initial step in the synthesis of an infective, lytic virus) sometimes results in the removal of some of the host cell DNA, which is packaged into defective bacteriophages; part of the bacteriophage DNA is removed and replaced at the other end by a gene of the host bacterium. Such a virus particle is called a transducing phage because, when it infects a bacterial cell, it can transmit the gene captured by λ phage DNA into the next bacterial cell it infects. Transduction by bacteriophages is an efficient means for transferring the genetic information of one bacterial cell to another.

This means of transferring genetic information, called lysogenic conversion, imparts genes with special functions to bacterial cells without such functions. It is common in bacteria and is an important aspect of the epidemiology (incidence, distribution, and control) of infectious diseases. For example, the bacterium Corynebacterium diphtheriae is the causative agent of diphtheria, but only when it contains the prophage of bacteriophage β, which codes for the toxin that is responsible for the disease.

Malignant Transformation

A phenomenon analogous to bacterial cell lysogeny occurs in animal cells infected with certain viruses. These animal viruses do not generally cause disease immediately for certain animal cells. Instead, animal cells are persistently infected with such viruses, the DNA of which (provirus) is integrated into the chromosomal DNA of the host cell. In general, cells with integrated proviral DNA are converted into cancer cells, a phenomenon known as malignant transformation. As is the case with bacterial prophages, the transformed animal cell contains no infectious virus but only the integrated provirus DNA, which replicates along with the dividing cell's chromosomes. Therefore, following mitosis of the transformed cell, each new cell receives a copy of the proviral DNA. The hallmark of these transformed animal cells is that their growth is uncontrollable; unlike normal cells, their growth is not inhibited by contact with other cells, and they lose their capacity to adhere (anchor) to certain surfaces. Growth of normal tissues and organs is also controlled by a genetic phenomenon called programmed cell death, or apoptosis, in which a certain number of cells will die and be eliminated after a finite number of divisions. Malignant transformation can impede programmed cell death, thus allowing the cells to grow uncontrolled and resulting in cancer.

Among the animal viruses that cause malignant transformation by integration of proviral DNA are several families of DNA viruses and one large family of RNA viruses, the Retroviridae. Viruses of the family Polyomaviridae, a group of papovaviruses, were perhaps the first to be associated with malignancy (causing death or illness) in animals. Polyomaviruses are widespread in mice; they can infect other rodents, and they can cause tumours in infected animals. Another virus of the family Polyomaviridae is simian virus 40 (SV40), originally isolated from cells of the African green monkey (Cercopithecus sabaeus), where it grows rapidly and kills the cells. Infection of rodent or human cells, however, results in an abortive infection (an incompatibility between the virus and the host cell) but sometimes induces malignancy (sarcomas or lymphomas) in the occasional cell that is transformed. Viruses related to polyomavirus and SV40 have been isolated from humans, one of which, the JC virus, appears to be the causative agent of a fatal neurological disease called progressive multifocal leukoencephalopathy. In general, however, the human papovaviruses are not clearly associated with disease.

Other papovaviruses include the papillomaviruses (family Papillomaviridae), which are also small polygonal viruses containing circular double-stranded DNA. The papillomaviruses are associated with usually benign (nonthreatening) but widespread tumours, called papillomas or polyps, occurring in human skin and the genital tract. Specific papillomaviruses have been identified in humans in common warts and in genital warts (condylomata acuminata). Cancers of the human genital tract, particularly uterine cancer of the cervix, are frequently found in association with human papillomavirus type 16 (HPV 16); the virus undoubtedly is transmitted as a venereal disease.

Certain viruses of the family Adenoviridae, originally found in the tonsils and adenoids of humans, cause malignant transformation in certain cells. This phenomenon of cancer induction under laboratory conditions has been studied widely, but there is no evidence that the common adenoviruses cause cancers in humans. The common viruses of the family Herpesviridae, however, including the common herpes simplex viruses that cause cold sores and the venereal disease genital herpes, are suspected of being causative agents of cancer. Like the adenoviruses, the herpesviruses can cause malignant transformations, and their DNA is integrated into the host cell chromosome. A herpesvirus known as the Epstein-Barr virus causes a frequently fatal childhood cancer called Burkitt lymphoma as well as the nonmalignant disease infectious mononucleosis. The herpesvirus cytomegalovirus lies dormant in the tissues of most humans and can be induced to cause fatal diseases in infants and immunocompromised adults. A different herpesvirus causes chickenpox (varicella); the same virus lies latent in the tissues for long periods of time (perhaps years or decades) and later undergoes recrudescence (the recurrence of symptoms after they have abated) to cause the painful skin and neurological disease called herpes zoster, or shingles. In addition, there are herpesviruses that cause disease in animals—for example, the widespread and usually fatal disease in chickens called Marek's disease. The widespread distribution of viruses of the family Herpesviridae is evident from other diseases in monkeys and frogs.

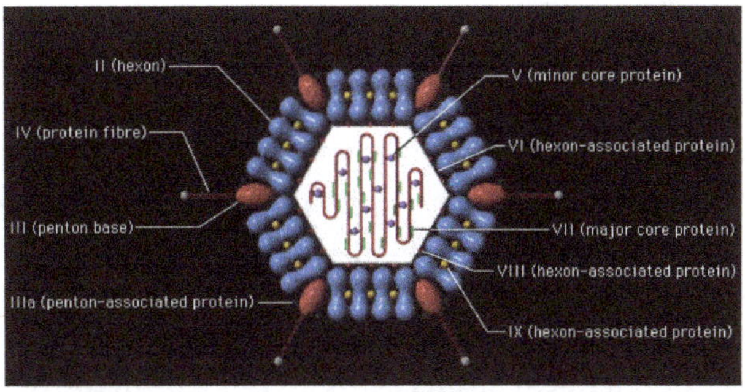

The structure of an adenovirus showing its components and the polypeptides (proteins) exhibited by staining after electrophoresis.

The viruses of the family Retroviridae are perhaps the most widely distributed of the transforming viruses that infect eukaryotic cells ranging from yeast to humans. It was suggested early in the 20th century that viruses cause leukemias and lymphomas in birds. In 1911 the American pathologist Peyton Rous first described a virus that causes sarcomas in chickens.

The virions of retroviruses are spherical (or polygonal) and are surrounded by a lipid membrane containing a glycoprotein that recognizes and binds to cell receptors of a particular species (type-specific glycoproteins). Retrovirus genomes consist of two identical RNA molecules, each with 7,000 to 10,000 nucleotides. Associated with the virion RNA is an enzyme, an RNA-dependent DNA polymerase, also called a reverse transcriptase. Using the virion RNA as a template, the reverse transcriptase catalyzes the synthesis of a linear DNA molecule complementary to the virion RNA. The new complementary strand of DNA also serves as a template for the reverse transcriptase, which makes a second anticomplementary DNA molecule, thus forming double-stranded DNA. The genomic RNA of fully infectious bird retroviruses, those that can replicate autonomously, has four genes that code sequentially for group-specific antigens, the reverse transcriptase, the envelope glycoprotein, and the sarcoma-transforming protein. At each end of the genome are homologous flanking nucleotide sequences, known as long terminal repeats (LTR), which code for double-stranded DNA that can recognize host cell DNA sequences for integration of the proviral DNA into the host cell chromosome. Many retroviruses are defective and cannot replicate in cells without helper (nondefective) retroviruses. The helper retroviruses generally transform fibroblastic cells, resulting in malignant sarcomas, whereas the defective retroviruses transform blood-cell precursors, resulting in leukemias.

Many different retroviruses have been identified as causative agents of cancers in birds, rodents (particularly mice), domestic cats, monkeys, and humans. Certain lymphatic leukemias in humans are caused by human T-cell leukemia virus (HTLV); acquired immune deficiency syndrome (AIDS) is caused by a retrovirus called human immunodeficiency virus (HIV).

Retroviruses originated from genes in many different species of animals and even lower forms of life. Individual retroviruses are limited in their host range and do not readily cross species barriers. Virtually every retrovirus studied to date is analogous to the genes normally found in animals (including humans), known as proto-oncogenes, genes that are involved with regulating normal cell growth and development and that also have the potential to change into cancer-causing genes. These proto-oncogenes have deoxynucleotide sequences closely, but not entirely, homologous (i.e., of the same type and order) to the nucleotide sequences of a corresponding viral cancer-causing gene, called an oncogene. Integration of retrovirus DNA into cell chromosomes results in cancer, but the proto-oncogenes do not become cancer-causing genes unless triggered by another event. Cancers caused by chemical or physical carcinogens in the environment probably often, if not invariably, are due to alterations in the sequences of proto-oncogenes that have converted them to oncogenes. Some of the DNA tumour viruses, such as SV40 or adenoviruses, may induce malignant transformation when their DNA is integrated in proximity to the site of a proto-oncogene. All cancers studied to date appear to be due to either mutations in proto-oncogenes or the inheritance of mutated tumour suppressor genes, which normally regulate the function of proto-oncogenes.

Disease

Although viruses were originally discovered and characterized on the basis of the diseases they cause, most viruses that infect bacteria, plants, and animals (including humans) do not cause disease. In fact, bacteriophages may be helpful in that they rapidly transfer genetic information from one bacterium to another, and viruses of plants and animals may convey genetic information between similar species, helping their hosts survive in hostile environments. In the future this could also be true for humans. Recombinant DNA biotechnology shows great promise for the repair of genetic defects. Afflicted persons are injected with cells transformed by viruses that carry a functional copy of the defective human gene. The virus integrates the normal gene into the DNA of the human cell.

Of those viruses that cause disease, some cause short-term (acute) diseases and others recurring or long-term (chronic) diseases. Some viruses cause acute disease from which there is fairly rapid recovery but may persist in the tissues, remaining dormant for long periods of time, and then become active again, bringing about serious disease decades later. Slowly progressive viruses have long incubation periods before the onset of disease. As mentioned above, the DNA of certain viruses becomes integrated into the genome of the host cell, often resulting in malignant transformation of cells, which become cancers.

The nature of the disease caused by a virus is generally a genetic property of the virus as well as of the host cells. Many viruses, however, can remain dormant in the tissues of the host (latency). Viruses that cause acute disease are generally, but not always, those that rapidly harm or destroy cells (cytopathic effects) and have the capacity to shut off protein or nucleic acid synthesis within the host cell.

Human poliovirus and related picornaviruses that infect other animal species are examples of acute infectious agents that shut down protein synthesis in the host cell soon after infection; these picornaviruses also inhibit cellular RNA and DNA synthesis. Another virus that rapidly kills the infected cell is the negative-strand vesicular stomatitis virus (VSV) of the family Rhabdoviridae; viral RNA newly synthesized by infectious VSV rapidly shuts off cellular RNA synthesis and, to a somewhat lesser extent, cellular protein synthesis. In both poliovirus and VSV, the infected cell dies within hours of the inhibition of cellular RNA and protein synthesis. Influenza A viruses of the family Orthomyxoviridae, which cause a highly contagious respiratory disease in humans, inhibit cellular macromolecular synthesis by several unique mechanisms, including blocking the maturation of cellular mRNAs and cleaving off the ends of cellular mRNAs in the nucleus of infected cells. Other viruses that inhibit cellular macromolecule synthesis and produce acute infections include the poxviruses, reoviruses, togaviruses, adenoviruses, and herpesviruses; the latter two persist in host tissues for long periods of time and cause chronic infection as well.

Many, if not most, diseases resulting from viral infection of vertebrates are caused not by a direct effect of the virus but rather by a secondary immune response. Essentially all viral proteins are recognized by vertebrate animals as immunologically foreign, and the immune systems of these animals mount two kinds of immune response, humoral and cellular. In humoral immunity, B lymphocytes, usually triggered by helper T lymphocytes, make antibodies (proteins that recognize and bind foreign molecules) to the viral protein. The antibody synthesized as a result of the immune response against a specific viral antigen usually benefits the infected host because that antibody can neutralize the infectivity of the specific virus in the blood and tissues of the infected host. Viruses inside the cell are not accessible to the antibody, because it cannot cross the cell membrane barrier.

In cellular immunity, a killer T cell recognizes and kills a virus-infected cell because of the viral antigen on its surface, thus aborting the infection because a virus will not grow within a dead cell. If the virus-infected cells are not essential for host functions, the killer T cell can prevent the spread of the infecting virus to other cells and distant tissues. Not infrequently, the virus-specific T lymphocyte kills vital cells such as nerve cells (neurons), muscle cells, and liver cells, all of which carry out important functions. In addition, the death of cells results in an inflammatory response, which also can damage vital tissues. Therefore, the cellular immune response to a viral infection can cause disease. In general, diseases caused by chronic viral infections, but also occasionally by subacute (between acute and chronic) viral infections, are caused by cellular immune responses that damage the virus-infected tissue.

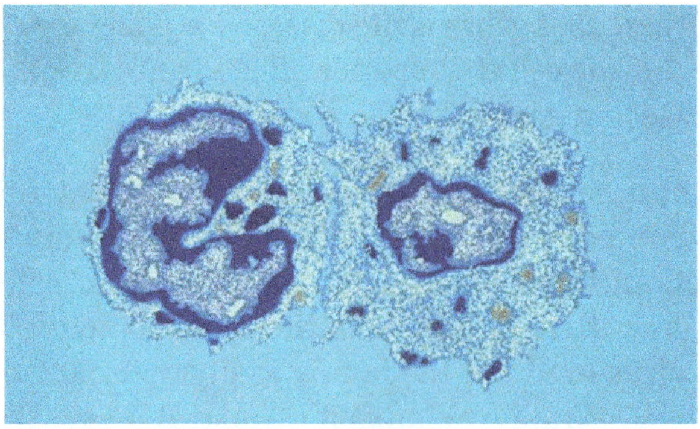

A cytotoxic T cell (left) recognizes antigens on the surface of a cell infected with a virus (right), enabling the T cell to bind to and kill the infected cell.

Infectious Patterns

Acute viral infections are of two types—local and systemic—both usually resulting from a direct effect of the invading virus on host tissue cells. Acute local infections generally occur at the site of viral infection. For example, acute respiratory infections include (1) the common cold, in which the rhinovirus infects only the nasal mucosa, (2) influenza, in which the virus is found in both nasal and bronchial mucosa, where severe damage can result in death, (3) flulike illnesses caused by adenoviruses localized in lymphoid tissue of the throat (although infection also can occur in the intestine and the eye or be spread to the heart), and (4) severe respiratory infections of infants and children, caused by parainfluenza viruses or respiratory syncytial viruses, which may be life-threatening. Examples of acute infections localized to the intestine include those that result in enteritis (bowel inflammation), which may be accompanied by diarrhea; these are often caused by rotaviruses and coronaviruses.

Many viruses transmitted by the respiratory route (from sneezes and coughs, for example) and limited to humans begin their cycle of infection in the upper respiratory tract (nose and throat) and then enter the bloodstream, where they are spread to distant tissues. Examples of such diseases are measles, mumps, and chickenpox, in which the growth of the specific virus in the mucosal cells of the throat during the first few days of infection usually results in mild fever and achiness; this stage is called the prodromal period of the illness. During the next few days, the virus enters the draining lymph nodes and then the bloodstream, where it is spread throughout the tissues of the body, resulting in fever and rash (in the case of measles and chickenpox) and inflammation of the parotid glands and, less frequently, the testes, ovaries, and joints (in the case of mumps). Varicella (chickenpox) virus rarely causes pneumonia, but all these viruses can cause meningitis and, rarely, encephalitis. A similar pattern of infection formerly occurred with smallpox, a disease that was more frequently fatal but now ostensibly has been eradicated.

A large number of viruses of the digestive tract (enteroviruses)—among them poliovirus, Coxsackie viruses, and echoviruses (enteric cytopathic human orphan virus)—also cause a two-phase illness. Enteroviruses grow initially in the intestinal tract and are transmitted by mouth through water, food, and other materials contaminated with feces. The viruses are resistant to the acid normally found in the stomach and thus reach the intestinal tract, where they multiply in living mucosal cells. This initial period of viral invasion and growth in the intestine causes either an initial mild febrile illness or is asymptomatic. Over the next few days these enteroviruses are spread from the intestinal mucosa to the draining lymph nodes, from which they invade the bloodstream, resulting in a condition known as viremia. From the bloodstream the viruses are widely spread to all tissues, but in most cases no symptomatic disease occurs. Poliovirus in less than 1 percent of cases affects the spinal cord or brain, resulting in paralysis or death. Different types of Coxsackie viruses and echoviruses can cause acute, usually nonfatal, illnesses such as meningitis, carditis, pleurisy, or rashes.

Many viral diseases are transmitted by bites of insects or other arthropods, and these infections usually begin in the skin or lymph nodes and rapidly invade the bloodstream. The nature of the disease caused by these arthropod-borne viruses (arboviruses) is determined by the affinity (tropism) of each virus for specific organs. Many that have an affinity for brain tissue cause encephalitis or meningitis, but others primarily infect the muscles, liver, heart, or kidneys. Virtually all these diseases are epidemic in character, and the viruses that cause them are the primary pathogens of birds and mammals.

The insect, usually a certain species of mosquito, takes a blood meal from the infected host bird or mammal and shortly thereafter bites a human, thus transmitting the virus. These arboviruses do not ordinarily multiply in the insect but simply reside on its proboscis. Examples of human epidemic diseases resulting from transmission of these often fatal arboviruses are encephalitis caused by viruses of the family Togaviridae and Flaviviridae, yellow fever and dengue caused by viruses of the family Flaviviridae, and hemorrhagic fevers caused by viruses of the families Bunyaviridae and Arenaviridae. Of considerable interest and concern is the identification of new strains of viruses, particularly a hantavirus of the Bunyaviridae family that was responsible for an epidemic in the early 1990s in the southwestern United States that resulted in considerable numbers of fatal human infections.

Latency

Inapparent infections (those that do not cause specific signs and symptoms) often result after exposure to picornaviruses, influenza viruses, rhinoviruses, herpesviruses, and adenoviruses but less frequently to measles and chickenpox viruses. In cases of inapparent infection, long-lasting immunity develops, but only to the strain of virus that has the same antigenic composition as the original infecting virus.

Certain of these viruses persist in the tissues of the host after the initial infection despite the presence of circulating antibodies to it in the blood and tissues. Such viruses probably reside inside cells, where they are protected from antibodies that cannot penetrate the cell membrane. Among persistent viruses are adenoviruses, measles virus, and, in particular, many kinds of herpesviruses. The genetic information of herpesviruses and adenoviruses can be integrated into the genome of the host cell, but it is believed that these viruses frequently, and the measles virus invariably, reside in cells in the form of extrachromosomal genes (genes not integrated in chromosomes). These dormant viruses can be activated by many factors, such as trauma, another infection, emotional stress, menstruation, excessive exposure to sunlight, and various illnesses.

The phenomenon of latency and reactivation is particularly common among viruses of the family Herpesviridae, which cause chronic or recurrent diseases: (1) herpes simplex virus type 1, which causes recurrent cold sores, (2) herpes simplex virus type 2 in genital tissue, which causes repeated herpetic infections of the vagina or penis, (3) cytomegalovirus, which usually produces an inapparent infection activated by simultaneously occurring disease to cause severe liver, lung, or nervous-system disease, and (4) varicella virus, which is the causative agent of chickenpox but which can be activated decades later to produce herpes zoster (shingles). A rare, but invariably fatal, disease of the nervous system is subacute sclerosing panencephalitis (SSPE), which is a progressive degenerative condition caused by measles virus (a paramyxovirus) lying dormant in brain cells for many years and then reactivated, usually in adolescence. There is no simple explanation for why latent viruses, such as those in the family Herpesviridae, that are present in the tissues of most adult humans can be activated to cause disease in some people but not in others.

Chronic and Slowly Progressive Diseases

Although some viruses multiply slowly, this is not always the explanation for the chronicity or the slow progression of the diseases caused by these viruses. Hepatitis, for example, is a subacute or chronic disease, with a long latent period, that is caused by at least five viruses with different properties. Hepatitis A is caused by a picornavirus usually transmitted by the fecal-oral route in a manner similar to that of poliovirus. Hepatitis B is caused by a small DNA virus that contains its own DNA polymerase and is transmitted by transfusion of blood and other blood products, by the sharing of nonsterile hypodermic needles among drug users, by sexual intercourse, or from mother to neonate. Hepatitis B virus is classified with similar viruses of birds in the family Hepadnaviridae. Most cases of hepatitis spread by the transfusion of blood or blood products or by needles shared by drug users are caused by a third, completely distinct virus—originally called non-A, non-B hepatitis but now known to be a member of the virus family Flaviviridae—designated hepatitis C virus. A fourth unique agent that causes hepatitis is designated hepatitis delta virus, which has not yet been classified taxonomically but is a small enveloped virus containing a circular RNA genome; hepatitis B virus serves as a helper for replication of hepatitis delta virus, the virions of which contain hepatitis B surface antigen (HBsAg). The fifth causative agent of viral hepatitis, largely occurring in Asia and Africa, is a small RNA virus tentatively classified as a member of the family Caliciviridae and designated hepatitis E virus.

Many other agents that appear to cause chronic and slowly progressive diseases, particularly those affecting the nervous system, have been identified. A fatal neurological disorder of sheep, called scrapie, has an incubation period of years and may be caused by a heat-resistant protein called a prion, which is self-replicating. Similar, rather obscure agents have been identified for two uncommon fatal disorders of the nervous system called Creutzfeldt-Jakob disease and kuru.

The disease now known as AIDS was first recognized in homosexuals and hemophiliacs about 1981 and continues to be disseminated throughout the world to become one of the most devastating epidemics of all time. AIDS is caused by HIV, a member of a genetically more complex group of the family Retroviridae called lentiviruses. Closely related viruses of monkeys and cats cause similar diseases. HIV is transmitted by blood and other body fluids and infects primarily helper T lymphocytes and other cells with CD4 surface receptors (cell surface proteins that react with antigens), to which the virus binds. After the virus has been dormant for years, destruction of T lymphocytes results in drastic depression of the immune system. Death almost invariably results from "opportunistic" infections such as pneumonia—caused by ordinarily nonpathogenic organisms such as Pneumocystis carinii—or tuberculosis or by cancers such as Kaposi sarcoma and lymphomas.

Prevention

The spread of many viral diseases can be prevented by hygienic factors such as efficient sanitation facilities, effective waste disposal, clean water, and personal cleanliness. Active immunization by vaccines (antigen-containing preparations that elicit the synthesis of antibodies and thus immunity) has been useful in preventing common epidemics caused by acutely infectious viruses.

The best example of such a preventable disease is smallpox, caused by a disease-producing virus that at one time was found worldwide. In 1796 the English physician Edward Jenner discovered that the milder cowpox virus could serve as a live vaccine (an antigenic preparation consisting of viruses whose disease-producing capacity has been weakened) for preventing smallpox; Jenner published his findings in 1798. The program of vaccination that resulted from Jenner's discovery is one of the greatest success stories in the annals of medicine; in 1980 the World Health Organization declared that the disease had been eliminated.

A different prospect is presented by rabies, an invariably fatal viral disease mentioned in ancient Greek literature. Transmitted by the bite of dogs and other domestic and wild animals, the rabies virus is more difficult to eradicate because it is present in wild animals throughout the world, except in certain island countries such as Great Britain and Australia. Influenza virus is also distributed worldwide, but, of the three major immunologic types, only one (type A) is responsible for large epidemics. The worldwide epidemic (pandemic) of influenza at the end of World War I is estimated to have caused 20 million deaths, mostly of adolescents and young adults. Because of virus mutations that produce minor antigenic changes every year and major antigenic shifts about every 10 years, influenza viruses have the capacity to resist inactivation by antibodies acquired by previous infection or vaccination. Development of effective vaccines to combat influenza is a difficult task, although existing vaccines are partially effective and are recommended for people at high risk—i.e., the elderly and those with chronic disease of the respiratory or circulatory systems.

Vaccines are most successful when directed against those viruses that do not mutate and that infect only humans. In addition to smallpox, a successful vaccine program has been carried out against polio. Polioviruses exist in only three antigenic types, each of which has not changed significantly for decades. The vaccines available are the "killed" (Salk) vaccine, composed of inactivated virus of the three types, and the "live" (Sabin) vaccine, composed of genetically attenuated viruses of the three types. In developed countries these vaccines, which were introduced in the 1950s, have lowered the incidence of paralysis resulting from polio. The disease still occurs in developing countries and recurs in some developed countries where vaccination programs have not been enforced. Rare cases of polio occur from the Sabin vaccine strain of type-3 poliovirus, which is genetically unstable and occasionally reverts to the virulent form.

Vaccination can prevent diseases caused by strictly human viruses that exist in only one antigenic and stable type. Measles has been prevented in developed countries with routine vaccination. Measles, however, may still be the major cause of death in children in developing countries. Vaccination for mumps and chickenpox promises to be successful because the causative viruses of these diseases show little tendency to vary antigenically and are confined to humans. On the other hand, development of vaccines for the common cold caused by rhinoviruses, similar to polioviruses, will be a formidable, if not impossible, task because there are at least 100 antigenic types of the rhinovirus. Also daunting is the task of developing a vaccine against HIV. The major antigenic component of this virus is a surface-membrane-inserted glycoprotein (gp120), which has a startling rate of mutation. The extreme antigenic diversity that results from the mutability of the gene that codes for this protein would prevent HIV from being identified and attacked by circulating antibodies or killer T lymphocytes.

Treatment

Unlike bacteria, viruses mimic the metabolic functions of their host cells. Antibiotics are not effective against viruses. It is difficult to identify chemical compounds that inhibit the multiplication of viruses but do not slow the functions of, or are not toxic to, the host cell. Despite this difficulty, an effective antiviral drug has been developed against influenza virus. This drug targets a viral enzyme called the neuraminidase and is orders of magnitude less active against nonviral neuraminidases. These neuraminidase inhibitors are most effective when administered prophylactically or within the first 30 hours of symptom onset and can be used to limit the spread of influenza virus and to complement the administration of vaccines. Other chemicals that exert a selectively greater effect on viral replication than they do on cell replication include ribavirin, acyclovir, and zidovudine (azidothymidine [AZT]). These drugs have been partially effective in improving, if not curing, viral diseases without causing major toxic side effects. AZT has been used with some success in prolonging the lives of patients with AIDS.

Certain natural products of cells, called interferons, may have potential antiviral and anticancer properties. Interferons are proteins normally synthesized by the cells of vertebrates, including humans, either intrinsically and without stimulation or in response to certain viral infections, chemicals, or immune reactions. In general, the multiplication of viruses is inhibited by interferons, some to a much greater extent than others. Interferons are generally species-specific; i.e., they are effective in inhibiting viral infection only in cells of the same species that naturally synthesize the interferon.

There are three classes of interferons: α-interferons, produced by blood leukocytes; β-interferons, produced by tissue cells and fibroblasts; and γ-interferons (also called immune interferons or interleukins), produced by immune reactions in blood lymphocytes. Interferons are now known to be a subset of a large group of natural cellular substances called cytokines, which signal cells to perform specific functions. Until recently, interferons were difficult to produce commercially because cells and tissues synthesize only small amounts of them. Through recombinant DNA technology, however, large amounts of interferon can be produced.

There has been some success in using interferons to treat viral diseases, such as colds caused by rhinoviruses, infections caused by herpesviruses, and benign tumours and warts caused by papillomaviruses. Local administration at the sites of viral infection affords the best results, although injections of large amounts of interferons can be harmful, probably because they tend to inhibit protein synthesis in the host cell.

Classification

Distinguishing Taxonomic Features

Viruses are classified on the basis of their nucleic acid content, their size, the shape of their protein capsid, and the presence of a surrounding lipoprotein envelope.

The primary taxonomic division is into two classes based on nucleic acid content: DNA viruses or RNA viruses. The DNA viruses are subdivided into those that contain either double-stranded or single-stranded DNA. The RNA viruses also are divided into those that contain double-stranded or single-stranded RNA. Further subdivision of the RNA viruses is based on whether the RNA genome is segmented. If the viruses contain single-stranded RNA as their genetic information, they are divided into positive-strand viruses if the RNA is of messenger sense (directly translatable into proteins) or negative-strand viruses if the RNA must be transcribed by a polymerase into mRNA.

All viruses falling into one of these nucleic acid classifications are further subdivided on the basis of whether the nucleocapsid (protein coat and enclosed nucleic acid) assumes a rodlike or a polygonal (usually icosahedral) shape. The icosahedral viruses are further subdivided into families on the basis of the number of capsomeres making up the capsids. Finally, all viruses fall into two classes depending on whether the nucleocapsid is surrounded by a lipoprotein envelope.

Some virologists adhere to a division of viruses into those that infect bacteria, plants, or animals; these classifications have some validity, particularly for the unique bacterial viruses with tails, but there is otherwise so much overlap that taxonomy based on hosts seems unworkable. Classification based on diseases caused by viruses also is not tenable, because closely related viruses frequently do not cause the same disease. Eventually, it is likely that the classification of viruses will be based on their nucleotide sequences and their mode of replication rather than on structural components, as is now the case.

The basic taxonomic group is called a family, designated by the suffix -viridae. The major taxonomic disagreement among virologists is whether to segregate viruses within a family into a specific genus and further subdivide them into species names. In the first decade of the 21st century, there occurred a shift toward the use of binomial nomenclature, dividing viruses into italicized genera and species. This move was prompted in large part by the International Committee on Taxonomy of Viruses (ICTV), a member group of the International Union of Microbiological Societies. The ICTV oversees the ongoing process of devising and maintaining a universal classification scheme for viruses. In the virus classification hierarchy, the ICTV recognizes orders, families, subfamilies, genera, and species. The placement of viruses in these groups is based on information provided by study groups composed of experts on specific types of viruses.

In the ICTV system, each species of virus is generally recognized as representing a group of isolates, or viruses with distinct nucleic acid sequences. Thus, a single species of virus may sometimes contain more than one isolate. Although the isolates of a species possess unique genetic sequences, they all descend from the same replicating lineage and therefore share particular genetic traits. Furthermore, isolates of a species also share in common the ability to thrive within a specific ecological niche. As scientists identify new isolates and species, the classification of viruses is expected to become increasingly complex. The following scheme presents examples of well-characterized DNA and RNA viruses as they are classified on the basis of the ICTV system.

References

- Microorganism, biologydictionary.net, Retrieved 12 July, 2020
- Bonachela, juan; raghib, michael; levin, simon (feb 21, 2012). “dynamic model of flexible phytoplankton nutrient uptake”. Pnas. 108 (51): 20633–20638. Doi:10.1073/pnas.1118012108. Pmc 3251133. Pmid 22143781.
- Bacteria, science: britannica.com, Retrieved 13 August, 2020
- Starckx, senne (31 october 2012) a place in the sun - algae is the crop of the future, according to researchers in geel flanders today, Retrieved 8 december 2020
- Fungus, science: britannica.com, Retrieved 14 January, 2020
- Bonachela, juan; raghib, michael; levin, simon (feb 21, 2012). “dynamic model of flexible phytoplankton nutrient uptake”. Pnas. 108 (51): 20633–20638. Doi:10.1073/pnas.1118012108. Pmc 3251133. Pmid 22143781
- Protozoan, science: britannica.com, Retrieved 15 February, 2020
- Virus, science: britannica.com, Retrieved 16 March, 2020

Fundamental Concepts in Microbiology

CHAPTER 4

The fundamental concepts of microbiology include sterilization and disinfection, microbiological culture, microbial metabolism, microbial cooperation and microbial loop. The chapter closely examines these key concepts of microbiology to provide an extensive understanding of the subject.

Cell morphology is extremely diverse. Specific shapes are the consequence of adaptive pressures optimizing bacterial fitness. Shape affects critical biological functions, including nutrient acquisition, motility, dispersion, stress resistance and interactions with other organisms. Although the characteristic shape of a bacterial species remains unchanged for vast numbers of generations, periodical variations occur throughout the cell (division) and life cycles, and these variations can be influenced by environmental conditions. Bacterial morphology is ultimately dictated by the net-like peptidoglycan (PG) sacculus. The species-specific shape of the PG sacculus at any time in the cell cycle is the product of multiple determinants. Some morphological determinants act as a cytoskeleton to guide biosynthetic complexes spatiotemporally, whereas others modify the PG sacculus after biosynthesis. Accumulating evidence supports critical roles of morphogenetic processes in bacteria-host interactions, including pathogenesis.

The variation of bacterial cell shapes is often underappreciated. In addition to the well-known rods and cocci, more exotic shapes such as stars, mustaches, serpentines, and branches represent a large, although undefined, proportion. The characteristic morphology of a bacterial species is maintained through countless generations but is periodically modified within set limits during bacterial division and life cycles. Bacterial shape is genetically determined, but physical forces (internal and external) exerted on cells are increasingly recognized as major players in morphogenesis. To ensure constant bacterial morphology over generations despite these forces, shape maintenance must be an active process guided by robust regulatory circuits. This is evidenced by the development of aberrant morphology upon mutations. Shape dictates the interactions between a bacterial cell and its environment, most notably small-molecule traffic (via the surface/volume ratio), motility, formation of multicellular aggregates, habitat colonization (including eukaryotic hosts and consequently pathogenesis and symbiosis), predation, and resistance. Therefore, morphogenesis should be viewed as a major evolutionary and adaptive process that contributes greatly to prokaryotic ubiquity and versatility.

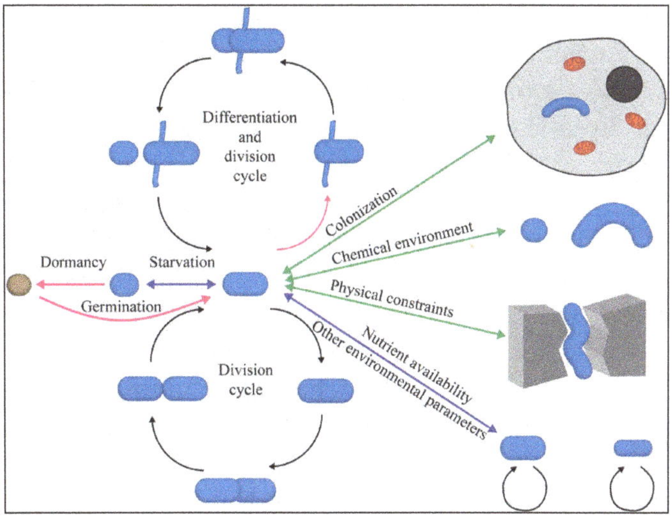

Morphological plasticity and the bacterial life cycle. The scheme illustrates the continuous modulations affecting bacterial shape throughout the life cycle. These changes can be either cyclic (division cycles) or sporadic in response to changing conditions, the presence of chemicals, colonization of other organisms or environments, nutrient depletion or abundance, etc. Most shape alterations are reversible (double-headed arrows) and could be considered adaptive phenomena, whereas others are irreversible (single-headed magenta arrows) and represent bona fide morphological differentiation processes, such as sporulation (orange sphere) or polymorphic cell cycles.

Bacterial shape is primarily dictated by the peptidoglycan (PG) sacculus, a polymeric macromolecular structure that surrounds the cytoplasmic membrane and is the only "solid" element in the bacterial envelope. PG is present in essentially all bacteria the number of exceptions is quickly dwindling as better detection methods are developed and wraps the cytoplasmic membrane like an elastic net. PG is a polymer of glycan chains crosslinked by peptides. The structure of the monomeric subunit, N-acetyl-glucosaminyl-N-acetyl-muramyl-L-alanyl-D-glutaminyl-L-(meso)diaminopimelyl-D-alanyl-D-alanine (GlcNAc-MurNAc-L-Ala-D-Glu-L-mesoDAP-D-Ala-D-Ala), is remarkably conserved throughout the bacterial phylogenetic tree. The few variations are either a change in the amino acid sequence of the stem peptide (almost always the di-amino acid at position 3) or the consequence of accessory reactions that modify the basic subunit (e.g., O-acetylation of sugars or amidation of dicarboxylic amino acids).

The precursors for PG biosynthesis, uridine diphosphate-N-acetylglucosamine (UDP-GlcNAc) and UDP-MurNAc-pentapeptide, are synthesized in the cytoplasm by the enzymes MurA-F. The enzyme MraY couples the UDP-MurNAc-pentapeptide to undecaprenyl phosphate to produce the membrane-anchored lipid I. Subsequent addition of GlcNAc to lipid I by MurG results in inward-oriented lipid II molecules. Translocation to the outer face of the cytoplasmic membrane is performed by the flippase MurJ, with the likely participation in some species of AmiJ and the SEDS (shape, elongation, division and sporulation) proteins RodA and FtsW. Once transferred to the external side of the cytoplasmic membrane, the GlcNAc-MurNAc-pentapeptide moiety of lipid II becomes accessible to enzymes with glycosyltransferase (GT) and transpeptidase (TP) activities, which catalyze linear polymerization and peptide crosslinking, respectively. The undecaprenyl diphosphate released in the polymerization reaction is flipped back, dephosphorylated and reused for the cyclical transport of new precursors. Bifunctional proteins with GT and TP activities are universal and concurrent with monofunctional representatives of both activities. The SEDS protein RodA was recently identified as a novel GT enzyme in Bacillus subtilis, and seems to play the same role in Escherichia coli. A ubiquitous class of enzymes involved in crosslinking is DD-transpeptidases, which are inhibited by covalent binding of beta-lactams and accordingly were first identified as penicillin-binding proteins (PBPs). In mature PG, D,D-crosslinks between the D-Ala at position 4 in the stem peptide of one subunit and the di-amino acid at position 3 (either directly or through intermediate peptides) of a nearby stem peptide are universal. Additional crosslinking mechanisms involving specific sets of enzymes and distinctive stereochemistry are relatively common.

As the PG layer is a covalently closed structure, the addition of new material requires concomitant cleavage of pre-existing bonds by PG hydrolases to permit enlargement of the sacculus. PG remodeling and maturation are mostly mediated by PG hydrolases. As a group, these enzymes target every bond (glycosidic and peptidic) sustaining the PG fabric. Organisms can encode many hydrolases, which are often redundant (35 and counting in E. coli). In addition to the enlargement process, the sacculus is subject to a complex and dynamic metabolism involving a large number of proteins that are not directly involved in the integration of precursors. Modifications related to subunit aging, growth state, nutritional conditions, population density, and stress response have been reported. Of particular interest to the present work are modifications causing changes in shape and the differentiation of new structures during the cell and life cycles of bacteria.

Generation of Cell Shape

Because of its covalently closed, net-like structure, the PG sacculus retains a specific shape and imposes this shape on the bacterial cell body. Despite a certain degree of deformability due to the elastic nature of the PG fabric, isolated sacculi faithfully retain the shape of the corresponding cell. However, the sacculi themselves lack the inbuilt information and/or features to determine their precise shape. Coding of epigenetic structural information in the 3D organization of the molecule has been proposed. However, no hard evidence supports this hypothesis. Furthermore, the current view of the sacculus as a relatively disordered array and the ability of "cell wall-less" forms to regenerate bacillary shapes argue against such coding. Even if a particular disposition of incoming new precursors might be favored by the pre-existing order (or lack thereof) of the older material, this does not necessarily indicate a global shape-defining role. Therefore, it seems reasonable to assume that while the sacculus is the element that confers and preserves a defined shape and size, the generation of that shape depends on the dynamics and topology of biosynthetic complexes rather than the sacculus itself.

The simple growth of a closed net subjected to cytoplasmic turgor pressure poses some critical constraints on the incorporation of new material to cause an effective enlargement. These constraints must be overcome by morphogenetic mechanisms. Simple attachment of incoming precursors to the sacculus would result only in thickening. Indeed, PG endopeptidases that permit expansion by cleaving existing crosslinked peptides have been identified in E. coli, Vibrio cholerae, and B. subtilis. However, if insertion of new material, and concomitant cleavage of old crosslinks, would happen constantly and evenly over the whole surface of the sacculus, this would lead to a homogeneous expansion of the growing structure. This mechanism by itself would not allow for the differentiation of new features. To generate shapes other than a sphere, incorporation must occur at distinct rates in different locations and for defined periods of time. Budding, for instance, would require a faster rate of precursor incorporation at the budding site than in the surrounding area. The morphogenetic process in bacteria not only requires physical enlargement, but also must allow periodic division events to increase the number of individuals.

As the mode of division of common model organisms, symmetrical binary fission is the best-known division mechanism and represents an elegant, intuitive mechanism to ensure shape conservation. However, alternative ways of division also occur. The only critical condition for division is the equitable distribution of both the genetic material and the biochemical components required to express the genetic potential. Division must be regulated in such a way that further divisions are not allowed before these conditions are fulfilled by the daughter cells. Many bacterial species divide by alternative mechanisms, often producing offspring cells that are quite dissimilar in size, shape and physiology from the mother cells. In these instances, the "juvenile" cells must undergo complex developmental programs to generate the characteristic morphology before committing to a subsequent round of division.

Cytokinesis implies the scission of the bacterial cell wall at genetically determined locations and cell cycle times while preserving cell integrity. The sacculus is a common substrate in cytokinesis and growth (enlargement and differentiation), which are mediated by closely related enzymatic complexes.

STERILIZATION AND DISINFECTION

Sterilization is defined as the process where all the living microorganisms, including bacterial spores are killed. Sterilization can be achieved by physical, chemical and physiochemical means. Chemicals used as sterilizing agents are called chemisterilants.

Disinfection is the process of elimination of most pathogenic microorganisms (excluding bacterial spores) on inanimate objects. Disinfection can be achieved by physical or chemical methods. Chemicals used in disinfection are called disinfectants. Different disinfectants have different target ranges, not all disinfectants can kill all microorganisms. Some methods of disinfection such as filtration do not kill bacteria, they separate them out. Sterilization is an absolute condition while disinfection is not. The two are not synonymous.

Decontamination is the process of removal of contaminating pathogenic microorganisms from the articles by a process of sterilization or disinfection. It is the use of physical or chemical means to remove, inactivate, or destroy living organisms on a surface so that the organisms are no longer infectious.

Sanitization is the process of chemical or mechanical cleansing, applicable in public health systems. Usually used by the food industry. It reduces microbes on eating utensils to safe, acceptable levels for public health.

Asepsis is the employment of techniques (such as usage of gloves, air filters, uv rays etc) to achieve microbe-free environment.

Antisepsis is the use of chemicals (antiseptics) to make skin or mucus membranes devoid of pathogenic microorganisms.

Bacteriostasis is a condition where the multiplication of the bacteria is inhibited without killing them.

Bactericidal is that chemical that can kill or inactivate bacteria. Such chemicals may be called variously depending on the spectrum of activity, such as bactericidal, virucidal, fungicidal, microbicidal, sporicidal, tuberculocidal or germicidal.

Antibiotics are substances produced by one microbe that inhibits or kills another microbe. Often the term is used more generally to include synthetic and semi-synthetic antimicrobial agents.

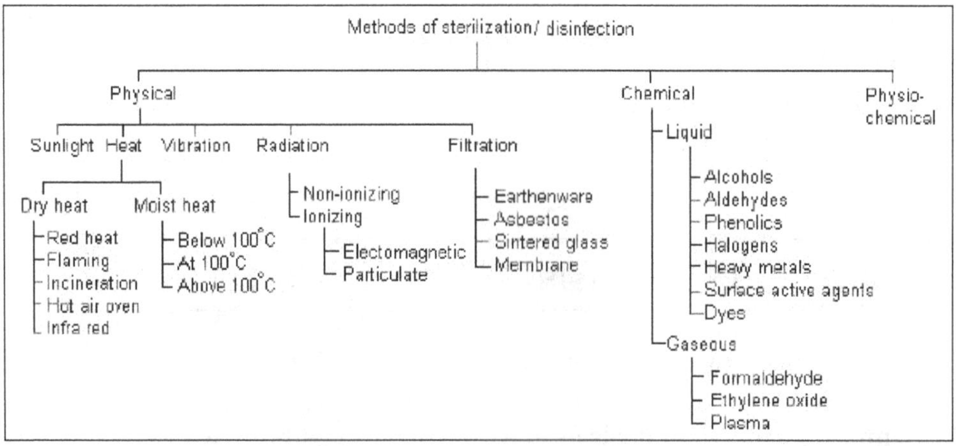

Physical Methods of Sterilization

Sunlight: The microbicidal activity of sunlight is mainly due to the presence of ultra violet rays in it. It is responsible for spontaneous sterilization in natural conditions. In tropical countries, the sunlight is more effective in killing germs due to combination of ultraviolet rays and heat. By killing bacteria suspended in water, sunlight provides natural method of disinfection of water bodies such as tanks and lakes. Sunlight is not sporicidal, hence it does not sterilize. Heat: Heat is considered to be most reliable method of sterilization of articles that can withstand heat. Heat acts by oxidative effects as well as denaturation and coagulation of proteins. Those articles that cannot withstand high temperatures can still be sterilized at lower temperature by prolonging the duration of exposure.

Factors affecting Sterilization by Heat

- Nature of heat: Moist heat is more effective than dry heat.
- Temperature and time: Temperature and time are inversely proportional. As temperature increases the time taken decreases.
- Number of microorganisms: More the number of microorganisms, higher the temperature or longer the duration required.
- Nature of microorganism: Depends on species and strain of microorganism, sensitivity to heat may vary. Spores are highly resistant to heat.
- Type of material: Articles that are heavily contaminated require higher temperature or prolonged exposure. Certain heat sensitive articles must be sterilized at lower temperature.
- Presence of organic material: Organic materials such as protein, sugars, oils and fats increase the time required.

Action of Heat

Dry heat acts by protein denaturation, oxidative damage and toxic effects of elevated levels of electrolytes. The moist heat acts by coagulation and denaturation of proteins. Moist heat is superior to dry heat in action. Temperature required to kill microbe by dry heat is more than the moist heat. Thermal death time is the minimum time required to kill a suspension of organisms at a predetermined temperature in a specified environment.

Dry Heat

Red heat: Articles such as bacteriological loops, straight wires, tips of forceps and searing spatulas are sterilized by holding them in Bunsen flame till they become red hot. This is a simple method for effective sterilization of such articles, but is limited to those articles that can be heated to redness in flame.

Flaming: This is a method of passing the article over a Bunsen flame, but not heating it to redness. Articles such as scalpels, mouth of test tubes, flasks, glass slides and cover slips are passed through the flame a few times. Even though most vegetative cells are killed, there is no guarantee that spores too would die on such short exposure. This method too is limited to those articles that can be exposed to flame. Cracking of the glassware may occur.

Incineration: This is a method of destroying contaminated material by burning them in incinerator. Articles such as soiled dressings; animal carcasses, pathological material and bedding etc should be subjected to incineration. This technique results in the loss of the article, hence is suitable only for those articles that have to be disposed. Burning of polystyrene materials emits dense smoke, and hence they should not be incinerated.

Hot air oven: This method was introduced by Louis Pasteur. Articles to be sterilized are exposed to high temperature (160 °C) for duration of one hour in an electrically heated oven. Since air is poor conductor of heat, even distribution of heat throughout the chamber is achieved by a fan. The heat is transferred to the article by radiation, conduction and convection. The oven should be fitted with a thermostat control, temperature indicator, meshed shelves and must have adequate insulation.

Articles sterilized: Metallic instruments (like forceps, scalpels, scissors), glasswares (such as petri-dishes, pipettes, flasks, all-glass syringes), swabs, oils, grease, petroleum jelly and some pharmaceutical products.

Sterilization process: Articles to be sterilized must be perfectly dry before placing them inside to avoid breakage. Articles must be placed at sufficient distance so as to allow free circulation of air in between. Mouths of flasks, test tubes and both ends of pipettes must be plugged with cotton wool. Articles such as petri dishes and pipettes may be arranged inside metal canisters and then placed. Individual glass articles must be wrapped in kraft paper or aluminum foils.

Sterilization cycle: This takes into consideration the time taken for the articles to reach the sterilizing temperature, maintenance of the sterilizing temperature for a defined period (holding time) and the time taken for the articles to cool down. Different temperature-time relations for holding time are 60 minutes at 160 °C, 40 minutes at 170 °C and 20 minutes at 180 °C. Increasing temperature by 10 degrees shortens the sterilizing time by 50 percent. The hot air oven must not be opened until the temperature inside has fallen below 60 °C to prevent breakage of glasswares.

Sterilization control: Three methods exist to check the efficacy of sterilization process, namely physical, chemical and biological.

- Physical: Temperature chart recorder and thermocouple.
- Chemical: Browne's tube No.3 (green spot, color changes from red to green).

- Biological: 10^6 spores of Bacillus subtilis var niger or Clostridium tetani on paper strips are placed inside envelopes and then placed inside the hot air oven. Upon completion of sterilization cycle, the strips are removed and inoculated into thioglycollate broth or cooked meat medium and incubated at 37 °C for 3-5 days. Proper sterilization should kill the spores and there should not be any growth.

Advantages: It is an effective method of sterilization of heat stable articles. The articles remain dry after sterilization. This is the only method of sterilizing oils and powders.

Disadvantages

- Since air is poor conductor of heat, hot air has poor penetration.
- Cotton wool and paper may get slightly charred.
- Glasses may become smoky.
- Takes longer time compared to autoclave.

Infra red rays: Infrared rays bring about sterilization by generation of heat. Articles to be sterilized are placed in a moving conveyer belt and passed through a tunnel that is heated by infrared radiators to a temperature of 180 °C. The articles are exposed to that temperature for a period of 7.5 minutes. Articles sterilized included metallic instruments and glassware. It is mainly used in central sterile supply department. It requires special equipments, hence is not applicable in diagnostic laboratory. Efficiency can be checked using Browne's tube No.4 (blue spot).

Moist Heat

Moist heat acts by coagulation and denaturation of proteins.

At Temperature below 100 °C

- Pasteurization: This process was originally employed by Louis Pasteur. Currently this procedure is employed in food and dairy industry. There are two methods of pasteurization, the holder method (heated at 63 °C for 30 minutes) and flash method (heated at 72 °C for 15 seconds) followed by quickly cooling to 13 °C. Other pasteurization methods include Ultra-High Temperature (UHT), 140 °C for 15 sec and 149 °C for 0.5 sec. This method is suitable to destroy most milk borne pathogens like Salmonella, Mycobacteria, Streptococci, Staphylococci and Brucella, however Coxiella may survive pasteurization. Efficacy is tested by phosphatase test and methylene blue test.
- Vaccine bath: The contaminating bacteria in a vaccine preparation can be inactivated by heating in a water bath at 60 °C for one hour. Only vegetative bacteria are killed and spores survive.

- Serum bath: The contaminating bacteria in a serum preparation can be inactivated by heating in a water bath at 56 °C for one hour on several successive days. Proteins in the serum will coagulate at higher temperature. Only vegetative bacteria are killed and spores survive.
- Inspissation: This is a technique to solidify as well as disinfect egg and serum containing media. The medium containing serum or egg are placed in the slopes of an inspissator and heated at 80-85 °C for 30 minutes on three successive days. On the first day, the vegetative bacteria would die and those spores that germinate by next day are then killed the following day. The process depends on germination of spores in between inspissation. If the spores fail to germinate then this technique cannot be considered sterilization.

At Temperature 100 °C

- Boiling: Boiling water (100 °C) kills most vegetative bacteria and viruses immediately. Certain bacterial toxins such as Staphylococcal enterotoxin are also heat resistant. Some bacterial spores are resistant to boiling and survive; hence this is not a substitute for sterilization. The killing activity can be enhanced by addition of 2% sodium bicarbonate. When absolute sterility is not required, certain metal articles and glasswares can be disinfected by placing them in boiling water for 10-20 minutes. The lid of the boiler must not be opened during the period.
- Steam at 100 °C: Instead of keeping the articles in boiling water, they are subjected to free steam at 100 °C. Traditionally Arnold's and Koch's steamers were used. An autoclave (with discharge tap open) can also serve the same purpose. A steamer is a metal cabinet with perforated trays to hold the articles and a conical lid. The bottom of steamer is filled with water and heated. The steam that is generated sterilizes the articles when exposed for a period of 90 minutes. Media such as TCBS, DCA and selenite broth are sterilized by steaming. Sugar and gelatin in medium may get decomposed on autoclaving, hence they are exposed to free steaming for 20 minutes for three successive days. This process is known as tyndallisation (after John Tyndall) or fractional sterilization or intermittent sterilization. The vegetative bacteria are killed in the first exposure and the spores that germinate by next day are killed in subsequent days. The success of process depends on the germination of spores.

At Temperature above 100 °C

Autoclave: Sterilization can be effectively achieved at a temperature above 100 °C using an autoclave. Water boils at 100 °C at atmospheric pressure, but if pressure is raised, the temperature at which the water boils also increases. In an autoclave the water is boiled in a closed chamber. As the pressure rises, the boiling point of water also raises. At a pressure of 15 lbs inside the autoclave, the temperature is said to be 121 °C. Exposure of articles to this temperature for 15 minutes sterilizes them. To destroy the infective agents associated with spongiform encephalopathies (prions), higher temperatures or longer times are used; 135 °C or 121 °C for at least one hour are recommended.

Advantages of steam: It has more penetrative power than dry air, it moistens the spores (moisture is essential for coagulation of proteins), condensation of steam on cooler surface releases latent heat, condensation of steam draws in fresh steam.

Different Types of Autoclave

Simple "pressure-cooker type" laboratory autoclave, Steam jacketed downward displacement laboratory autoclave and high pressure pre-vacuum autoclave.

Construction And Operation Of Autoclave:

A simple autoclave has vertical or horizontal cylindrical body with a heating element, a perforated try to keep the articles, a lid that can be fastened by screw clamps, a pressure gauge, a safety valve and a discharge tap. The articles to be sterilized must not be tightly packed. The screw caps and cotton plugs must be loosely fitted. The lid is closed but the discharge tap is kept open and the water is heated. As the water starts boiling, the steam drives air out of the discharge tap. When all the air is displaced and steam start appearing through the discharge tap, the tap is closed. The pressure inside is allowed to rise upto 15 lbs per square inch. At this pressure the articles are held for 15 minutes, after which the heating is stopped and the autoclave is allowed to cool. Once the pressure gauge shows the pressure equal to atmospheric pressure, the discharge tap is opened to let the air in. The lid is then opened and articles removed.

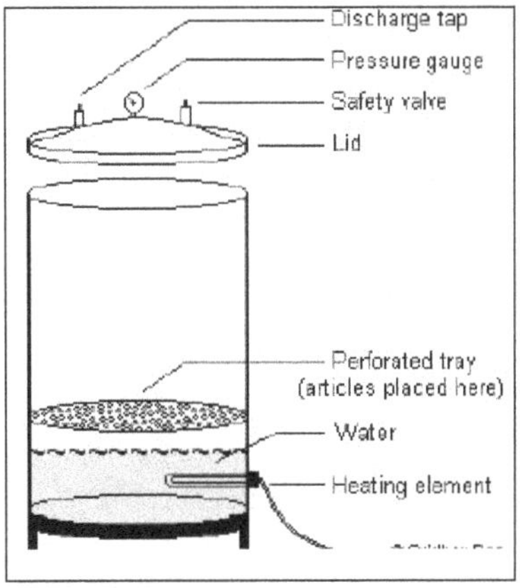

Articles sterilized: Culture media, dressings, certain equipment, linen etc.

Precautions: Articles should not be tightly packed, the autoclave must not be overloaded, air discharge must be complete and there should not be any residual air trapped inside, caps of bottles and flasks should not be tight, autoclave must not be opened until the pressure has fallen or else the contents will boil over, articles must be wrapped in paper to prevent drenching, bottles must not be overfilled.

Advantage: Very effective way of sterilization, quicker than hot air oven.

Disadvantages: Drenching and wetting or articles may occur, trapped air may reduce the efficacy, takes long time to cool.

Sterilization control: Physical method includes automatic process control, thermocouple and temperature chart recorder. Chemical method includes Browne's tube No.1 (black spot) and succinic acid (whose melting point is 121 °C) and Bowie Dick tape. Bowie Dick tape is applied to articles being autoclaved. If the process has been satisfactory, dark brown stripes will appear across the tape. Biological method includes a paper strip containing 10^6 spores of Geobacillus stearothermophilus.

Radiation

Two types of radiation are used, ionizing and non-ionizing. Non-ionizing rays are low energy rays with poor penetrative power while ionizing rays are high-energy rays with good penetrative power. Since radiation does not generate heat, it is termed "cold sterilization". In some parts of Europe, fruits and vegetables are irradiated to increase their shelf life up to 500 percent.

- Non-ionizing rays: Rays of wavelength longer than the visible light are non-ionizing. Microbicidal wavelength of UV rays lie in the range of 200-280 nm, with 260 nm being most effective. UV rays are generated using a high-pressure mercury vapor lamp. It is at this wavelength that the absorption by the microorganisms is at its maximum, which results in the germicidal effect. UV rays induce formation of thymine-thymine dimers, which ultimately inhibits DNA replication. UV readily induces mutations in cells irradiated with a non-lethal dose. Microorganisms such as bacteria, viruses, yeast, etc. that are exposed to the effective UV radiation are inactivated within seconds. Since UV rays don't kill spores, they are considered to be of use in surface disinfection. UV rays are employed to disinfect hospital wards, operation theatres, virus laboratories, corridors, etc. Disadvantages of using uv rays include low penetrative power, limited life of the uv bulb, some bacteria have DNA repair enzymes that can overcome damage caused by uv rays, organic matter and dust prevents its reach, rays are harmful to skin and eyes. It doesn't penetrate glass, paper or plastic.

- Ionizing rays: Ionizing rays are of two types, particulate and electromagnetic rays.

    - Electron beams are particulate in nature while gamma rays are electromagnetic in nature. High-speed electrons are produced by a linear accelerator from a heated cathode. Electron beams are employed to sterilize articles like syringes, gloves, dressing packs, foods and pharmaceuticals. Sterilization is accomplished in few seconds. Unlike electromagnetic rays, the instruments can be switched off. Disadvantage includes poor penetrative power and requirement of sophisticated equipment.

- Electromagnetic rays such as gamma rays emanate from nuclear disintegration of certain radioactive isotopes (Co60, Cs137). They have more penetrative power than electron beam but require longer time of exposure. These high-energy radiations damage the nucleic acid of the microorganism. A dosage of 2.5 megarads kills all bacteria, fungi, viruses and spores. It is used commercially to sterilize disposable petri dishes, plastic syringes, antibiotics, vitamins, hormones, glasswares and fabrics. Disadvantages include; unlike electron beams, they can't be switched off, glasswares tend to become brownish, loss of tensile strength in fabric. Gamma irradiation impairs the flavour of certain foods. Bacillus pumilus E601 is used to evaluate sterilization process.

Filtration

Filtration does not kill microbes, it separates them out. Membrane filters with pore sizes between 0.2-0.45 µm are commonly used to remove particles from solutions that can't be autoclaved. It is used to remove microbes from heat labile liquids such as serum, antibiotic solutions, sugar solutions, urea solution. Various applications of filtration include removing bacteria from ingredients of culture media, preparing suspensions of viruses and phages free of bacteria, measuring sizes of viruses, separating toxins from culture filtrates, counting bacteria, clarifying fluids and purifying hydatid fluid. Filtration is aided by using either positive or negative pressure using vacuum pumps. The older filters made of earthenware or asbestos are called depth filters.

Different Types of Filters Are:

- Earthenware filters: These filters are made up of diatomaceous earth or porcelain. They are usually baked into the shape of candle. Different types of earthenware filters are:
  - Pasteur-Chamberland filter: These candle filters are from France and are made up of porcelain (sand and kaolin). Similar filter from Britain is Doulton. Chamberland filters are made with various porosities, which are graded as L1, L1a, L2, L3, L5, L7, L9 and L11. Doulton filters are P2, P5 and P11.
  - Berkefeld filter: These are made of Kieselguhr, a fossilized diatomaceous earth found in Germany. They are available in three grades depending on their porosity (pore size); they are V (veil), N (normal) and W (wenig).

Quality of V grade filter is checked using culture suspension of Serrtia marcescens (0.75 μm).

- ◦ Mandler filter: This filter from America is made of kieselguhr, asbestos and plaster of Paris.

- Asbestos filters: These filters are made from chrysotile type of asbestos, chemically composed of magnesium silicate. They are pressed to form disc, which are to be used only once. The disc is held inside a metal mount, which is sterilized by autoclaving. They are available in following grades; HP/PYR (for removal of pyrogens), HP/EKS (for absolute sterility) and HP/EK (for claryfying).
- Sintered glass filters: These are made from finely ground glass that are fused sufficiently to make small particles adhere to each other. They are usually available in the form of disc fused into a glass funnel. Filters of Grade 5 have average pore diameter of 1-1.5 μm. They are washed in running water in reverse direction and cleaned with warm concentrated H2SO4 and sterilized by autoclaving.
- Membrane filters: These filters are made from a variety of polymeric materials such as cellulose nitrate, cellulose diacetate, polycarbonate and polyester. The older type of membrane, called gradocol (graded colloidion) membrane was composed of cellulose nitrate. Gradocol membranes have average pore diameter of 3-10 μm. The newer ones are composed of cellulose diacetate. These membranes have a pore diameter ranging from 0.015 μm to 12 μm. These filters are sterilized by autoclaving. Membrane filters are made in two ways, the capillary pore membranes have pores produced by radiation while the labyrinthine pore membranes are produced by forced evaporation of solvents from cellulose esters.

The disadvantages of depth filters are migration of filter material into the filtrate, absorption or retention of certain volume of liquid by the filters, pore sizes are not definite and viruses and mycoplasma could pass through. The advantages of membrane filters are known porosity, no retention of fluids, reusable after autoclaving and compatible with many chemicals. However, membrane filters have little loading capacity and are fragile.

Air Filters: Air can be filtered using HEPA (High Efficiency Particle Air) filters. They are usually used in biological safety cabinets. HEPA filters are at least 99.97% efficient for removing particles >0.3 μm in diameter. Examples of areas where HEPA filters are used include rooms housing severely neutropenic patients and those operating rooms designated for orthopedic implant procedures. HEPA filter efficiency is monitored with the dioctylphthalate (DOP) particle test using particles that are 0.3 μm in diameter.

Sonic and Ultrasonic Vibrations: Sound waves of frequency >20,000 cycle/second kills bacteria and some viruses on exposing for one hour. Microwaves are not particularly antimicrobial in themselves, rather the killing effect of microwaves are largely due to the heat that they generate. High frequency sound waves disrupt cells. They are used to clean and disinfect instruments as well as to reduce microbial load. This method is not reliable since many viruses and phages are not affected by these waves.

Chemical Methods of Disinfection

Disinfectants are those chemicals that destroy pathogenic bacteria from inanimate surfaces. Some chemical have very narrow spectrum of activity and some have very wide. Those chemicals that can sterilize are called chemisterilants. Those chemicals that can be safely applied over skin and mucus membranes are called antiseptics. An ideal antiseptic or disinfectant should have following properties:

- Should have wide spectrum of activity.
- Should be able to destroy microbes within practical period of time.
- Should be active in the presence of organic matter.
- Should make effective contact and be wettable.
- Should be active in any pH.
- Should be stable.
- Should have long shelf life.
- Should be speedy.
- Should have high penetrating power.
- Should be non-toxic, non-allergenic, non-irritative or non-corrosive.
- Should not have bad odour.
- Should not leave non-volatile residue or stain.
- Efficacy should not be lost on reasonable dilution.
- Should not be expensive and must be available easily.

Such an ideal disinfectant is not yet available. The level of disinfection achieved depends on contact time, temperature, type and concentration of the active ingredient, the presence of organic matter, the type and quantum of microbial load. The chemical disinfectants at working concentrations rapidly lose their strength on standing.

Classification of Disinfectants

- Based on consistency:
  - Liquid (E.g., Alcohols, Phenols).

- Gaseous (Formaldehyde vapor, Ethylene oxide).
- Based on spectrum of activity:
  - High level.
  - Intermediate level.
  - Low level.
- Based on mechanism of action:
  - Action on membrane (E.g., Alcohol, detergent).
  - Denaturation of cellular proteins (E.g., Alcohol, Phenol).
  - Oxidation of essential sulphydryl groups of enzymes (E.g., H_2O_2, Halogens).
  - Alkylation of amino-, carboxyl- and hydroxyl group (E.g., Ethylene Oxide, Formaldehyde).
  - Damage to nucleic acids (Ethylene Oxide, Formaldehyde).

	Vegetative cells	Mycobacteria	Spores	Fungi	Viruses	Examples
High level	+	+	+	+	+	Ethylene Oxide, Gluteral-dehyde, Formaldehyde
Intermediate level	+	+	-	+	+	Phenolics, halogens
Low level	+	-	-	+	+/-	Alcohols, quaternary ammonium compounds

Alcohols

Mode of action: Alcohols dehydrate cells, disrupt membranes and cause coagulation of protein.

Examples: Ethyl alcohol, isopropyl alcohol and methyl alcohol.

Application: A 70% aqueous solution is more effective at killing microbes than absolute alcohols. 70% ethyl alcohol (spirit) is used as antiseptic on skin. Isopropyl alcohol is preferred to ethanol. It can also be used to disinfect surfaces. It is used to disinfect clinical thermometers. Methyl alcohol kills fungal spores, hence is useful in disinfecting inoculation hoods.

Disadvantages: Skin irritant, volatile (evaporates rapidly), inflammable.

Aldehydes

Mode of action: Acts through alkylation of amino-, carboxyl- or hydroxyl group, and probably damages nucleic acids. It kills all microorganisms, including spores.

Examples: Formaldehyde, Gluteraldehyde.

Application: 40% Formaldehyde (formalin) is used for surface disinfection and fumigation of rooms, chambers, operation theatres, biological safety cabinets, wards, sick rooms etc. Fumigation is achieved by boiling formalin, heating paraformaldehyde or treating formalin with potassium permanganate. It also sterilizes bedding, furniture and books. 10% formalin with 0.5% tetraborate sterilizes clean metal instruments. 2% gluteraldehyde is used to sterilize thermometers, cystoscopes, bronchoscopes, centrifuges, anasethetic equipments etc. An exposure of at least 3 hours at alkaline pH is required for action by gluteraldehyde. 2% formaldehyde at 40oC for 20 minutes is used to disinfect wool and 0.25% at 60oC for six hours to disinfect animal hair and bristles.

Disadvantages: Vapors are irritating (must be neutralized by ammonia), has poor penetration, leaves non-volatile residue, activity is reduced in the presence of protein. Gluteraldehyde requires alkaline pH and only those articles that are wettable can be sterilized.

Phenol

Mode of action: Act by disruption of membranes, precipitation of proteins and inactivation of enzymes.

Examples: 5% phenol, 1-5% Cresol, 5% Lysol (a saponified cresol), hexachlorophene, chlorhexidine, chloroxylenol (Dettol).

Applications: Joseph Lister used it to prevent infection of surgical wounds. Phenols are coal-tar derivatives. They act as disinfectants at high concentration and as antiseptics at low concentrations. They are bactericidal, fungicidal, mycobactericidal but are inactive against spores and most viruses. They are not readily inactivated by organic matter. The corrosive phenolics are used for disinfection of ward floors, in discarding jars in laboratories and disinfection of bedpans. Chlorhexidine can be used in an isopropanol solution for skin disinfection, or as an aqueous solution for wound irrigation. It is often used as an antiseptic hand wash. 20% Chlorhexidine gluconate solution is used for pre-operative hand and skin preparation and for general skin disinfection. Chlorhexidine gluconate is also mixed with quaternary ammonium compounds such as cetrimide to get stronger and broader antimicrobial effects (eg. Savlon). Chloroxylenols are less irritant and can be used for topical purposes and are more effective against gram positive bacteria than gram negative bacteria. Hexachlorophene is chlorinated diphenyl and is much less irritant. It has marked effect over gram positive bacteria but poor effect over gram negative bacteria, mycobacteria, fungi and viruses. Triclosan is an organic phenyl ether with good activity against gram positive bacteria and effective to some extent against many gram negative bacteria including Pseudomonas. It also has fair activity on fungi and viruses.

Disadvantages: It is toxic, corrosive and skin irritant. Chlorhexidine is inactivated by anionic soaps. Chloroxylenol is inactivated by hard water.

Halogens

Mode of action: They are oxidizing agents and cause damage by oxidation of essential sulfydryl groups of enzymes. Chlorine reacts with water to form hypochlorous acid, which is microbicidal.

Examples: Chlorine compounds (chlorine, bleach, hypochlorite) and iodine compounds (tincture iodine, iodophores).

Applications: Tincture of iodine (2% iodine in 70% alcohol) is an antiseptic. Iodine can be combined with neutral carrier polymers such as polyvinylpyrrolidone to prepare iodophores such as povidone-iodine. Iodophores permit slow release and reduce the irritation of the antiseptic. For hand washing iodophores are diluted in 50% alcohol. 10% Povidone Iodine is used undiluted in pre and postoperative skin disinfection. Chlorine gas is used to bleach water. Household bleach can be used to disinfect floors. Household bleach used in a stock dilution of 1:10. In higher concentrations chlorine is used to disinfect swimming pools. 0.5% sodium hypochlorite is used in serology and virology. Used at a dilution of 1:10 in decontamination of spillage of infectious material. Mercuric chloride is used as a disinfectant.

Disadvantages: They are rapidly inactivated in the presence of organic matter. Iodine is corrosive and staining. Bleach solution is corrosive and will corrode stainless steel surfaces.

Heavy Metals

Mode of action: Act by precipitation of proteins and oxidation of sulfydryl groups. They are bacteriostatic. Examples: Mercuric chloride, silver nitrate, copper sulfate, organic mercury salts (e.g., mercurochrome, merthiolate)

Applications: 1% silver nitrate solution can be applied on eyes as treatment for opthalmia neonatorum (Crede's method). This procedure is no longer followed. Silver sulphadiazine is used topically to help to prevent colonization and infection of burn tissues. Mercurials are active against viruses at dilution of 1:500 to 1:1000. Merthiolate at a concentration of 1:10000 is used in preservation of serum. Copper salts are used as a fungicide.

Disadvantages: Mercuric chloride is highly toxic, are readily inactivated by organic matter.

Surface Active Agents

Mode of actions: They have the property of concentrating at interfaces between lipid containing membrane of bacterial cell and surrounding aqueous medium. These compounds have long chain hydrocarbons that are fat soluble and charged ions that are water-soluble. Since they contain both of these, they concentrate on the surface of membranes. They disrupt membrane resulting in leakage of cell constituents.

Examples: These are soaps or detergents. Detergents can be anionic or cationic. Detergents containing negatively charged long chain hydrocarbon are called anionic detergents. These include soaps and bile salts. If the fat-soluble part is made to have a positive charge by combining with a quaternary nitrogen atom, it is called cationic detergents. Cationic detergents are known as quaternary ammonium compounds (or quat). Cetrimide and benzalkonium chloride act as cationic detergents.

Application: They are active against vegetative cells, Mycobacteria and enveloped viruses. They are widely used as disinfectants at dilution of 1-2% for domestic use and in hospitals.

Disadvantages: Their activity is reduced by hard water, anionic detergents and organic matter. Pseudomonas can metabolise cetrimide, using them as a carbon, nitrogen and energy source.

Dyes

Mode of action: Acridine dyes are bactericidal because of their interaction with bacterial nucleic acids.

Examples: Aniline dyes such as crystal violet, malachite green and brilliant green. Acridine dyes such as acriflavin and aminacrine. Acriflavine is a mixture of proflavine and euflavine. Only euflavine has effective antimicrobial properties. A related dye, ethidium bromide, is also germicidal. It intercalates between base pairs in DNA. They are more effective against gram positive bacteria than gram negative bacteria and are more bacteriostatic in action.

Applications: They may be used topically as antiseptics to treat mild burns. They are used as paint on the skin to treat bacterial skin infections. The dyes are used as selective agents in certain selective media.

Hydrogen Peroxide

Mode of action: It acts on the microorganisms through its release of nascent oxygen. Hydrogen peroxide produces hydroxyl-free radical that damages proteins and DNA.

Application: It is used at 6% concentration to decontaminate the instruments, equipments such as ventilators. 3% Hydrogen Peroxide Solution is used for skin disinfection and deodorising wounds and ulcers. Strong solutions are sporicidal.

Disadvantages: Decomposes in light, broken down by catalase, proteinaceous organic matter drastically reduces its activity.

Ethylene OxiDE (EO)

Mode of action: It is an alkylating agent. It acts by alkylating sulfydryl-, amino-, carboxyl- and hydroxyl- groups. Properties: It is a cyclic molecule, which is a colorless liquid at room temperature. It has a sweet ethereal odor, readily polymerizes and is flammable.

Application: It is a highly effective chemisterilant, capable of killing spores rapidly. Since it is highly flammable, it is usually combined with CO_2 (10% CO2+ 90% EO) or dichlorodifluoromethane. It requires presence of humidity. It has good penetration and is well absorbed by porous material. It is used to sterilize heat labile articles such as bedding, textiles, rubber, plastics, syringes, disposable petri dishes, complex apparatus like heart-lung machine, respiratory and dental equipments.

Efficiency testing is done using Bacillus subtilis var niger.

Disadvantages: It is highly toxic, irritating to eyes, skin, highly flammable, mutagenic and carcinogenic.

Beta-propiolactone (BPL)

Mode of action: It is an alkylating agent and acts through alkylation of carboxyl- and hydroxyl- groups.

Properties: It is a colorless liquid with pungent to slightly sweetish smell. It is a condensation product of ketane with formaldehyde.

Application: It is an effective sporicidal agent, and has broad-spectrum activity. 0.2% is used to sterilize biological products. It is more efficient in fumigation that formaldehyde. It is used to sterilize vaccines, tissue grafts, surgical instruments and enzymes

Disadvantages: It has poor penetrating power and is a carcinogen.

Physio-chemical Method

Mode of action: A physio-chemical method adopts both physical and chemical method. Use of steam- formaldehyde is a physio-chemical method of sterilization, which takes into account action of steam as well as that of formaldehyde. Saturated steam at a pressure of 263 mm has a temperature of 70°C. The air is removed from the autoclave chamber and saturated steam at sub-atmospheric pressure is flushed in. Formaldehyde is then injected with steam in a series of pulses, each of 5-10 minutes. The articles are held at this holding temperature for one hour. Formaldehyde is then flushed by inflow of steam.

Disadvantages: Condensation of formaldehyde occurs and induction of large volume of formaldehyde wets the steam resulting in loss of latent heat.

Sterilization control: using paper strips containing 106 spores of G.stearothermophilus.

Testing of Disinfectants

A disinfectant must be tested to know the required effective dilution, the time taken to effect disinfection and to periodically monitor its activity. As disinfectants are known to lose their activity on standing as well as in the presence of organic matter, their activity must be periodically tested.

Different methods are:

- Koch's method.
- Rideal Walker Method.
- Chick Martin test.
- Capacity use dilution test (Kelsey-Sykes test).
- In-use test.

Koch's method: Spores of Bacillus anthracis were dried on silk thread and were subjected to action of disinfectants. Later, it was washed and transferred to solid medium.

Rideal Walker method: This method relies on the estimation of phenol coefficient. Phenol coefficient of a disinfectant is calculated by dividing the dilution of test disinfectant by the dilution of phenol that disinfects under predetermined conditions. Both the phenol and the test disinfectant are diluted from 1/95 to 1/115 and their bactericidal activity is determined against Salmonella typhi suspension. Subcultures are performed from both the test and phenol at intervals of 2.5, 5, 7.5 and 10 minutes. The plates are incubated for 48-72 hours at 37°C. That dilution of disinfectant which disinfects the suspension in a given time is divided by that dilution of phenol which disinfects the suspension in same time gives its phenol coefficient.

Disadvantages of the Rideal-Walker test are: No organic matter is included; the microorganism Salmonella typhi may not be appropriate; the time allowed for disinfection is short; it should be used to evaluate phenolic type disinfectants only.

Chick Martin test: This test also determines the phenol coefficient of the test disinfectant. Unlike in Rideal Walker method where the test is carried out in water, the disinfectants are made to act in the presence of yeast suspension (or 3% dried human feces). Time for subculture is fixed at 30 minutes and the organism used to test efficacy is S.typhi as well as S.aureus. The phenol coefficient is lower than that given by Rideal Walker method.

	Rideal -Walker	Chick-Martin
Volume medium	5.0 ml	10.0 ml
Diluent for test disinfectant	Water	Yeast suspension

Reaction temperature	17.5±0.5°C	30°C
Organism	Salmonella typhi	Salmonella typhi, Staphylococcus aureus
Sampling times	2.5, 5.0, 7.5, 10.0 min.	30.0 min.
Calculation of coefficient	Dilution test killing in 7.5 min divided by same for phenol	Mean concentration of phenol showing no growth after 30 min. divided by same for test

The classical tests such as Rideal - Walker or Chick - Martin are not practicable.

Capacity use Dilution Test

Inoculum of four different test organisms, namely Staphylococcus aureus, Escherichia coli, Pseudomonas aeruginosa and Proteus vulgaris are added to the disinfectant in three successive. Dried yeast is included to simulate presence of organic matter. The method can be carried out under 'clean' or 'dirty' conditions. The dilutions of the disinfectant are made in hard water for clean conditions and in yeast suspension for dirty conditions. Test organism alone or with yeast is added at 0,10 and 20 minutes interval. The contact time of disinfectant and test organism is 8 min. The disinfectant is evaluated on its ability to kill microorganisms or lack of it and the result is reported as a pass or a fail and not as a coefficient. The capacity test of Kelsey and Sykes gives a good guideline for the dilution of the preparation to be used. Disadvantage of this test is the fact that it is rather complicated.

In-use Test

The routine monitoring of disinfectant in use can be done by the 'in use' test of Maurer. This test is intended to estimate the number of living organism in a vessel of disinfectant in actual use.

The disinfectant that is already in use is diluted 1 in 10 by mixing 1 ml of the disinfectant with 9 ml of sterile nutrient broth. Ten drops of the diluted disinfectant (each 0.02 ml) is placed on two nutrient agar plates. One plate is incubated at 37oC for 3 days while the other is held at room temperature for 7 days. The number of drops that yielded growth is counted after incubation. If there growth in more than five drops on either plate, it represents failure of disinfectant.

MICROBIOLOGICAL CULTURE

A microbiological culture, or microbial culture, is a method of multiplying microbial organisms by letting them reproduce in predetermined culture medium under controlled laboratory conditions. Microbial cultures are foundational and basic diagnostic methods used extensively as a research tool in molecular biology.

Microbial cultures are used to determine the type of organism, its abundance in the sample being tested, or both. It is one of the primary diagnostic methods of microbiology and used as a tool to determine the cause of infectious disease by letting the agent multiply in a predetermined medium. For example, a throat culture is taken by scraping the lining of tissue in the back of the throat and blotting the sample into a medium to be able to screen for harmful microorganisms, such as *Streptococcus pyogenes*, the causative agent of strep throat. Furthermore, the term culture is more generally used informally to refer to "selectively growing" a specific kind of microorganism in the lab.

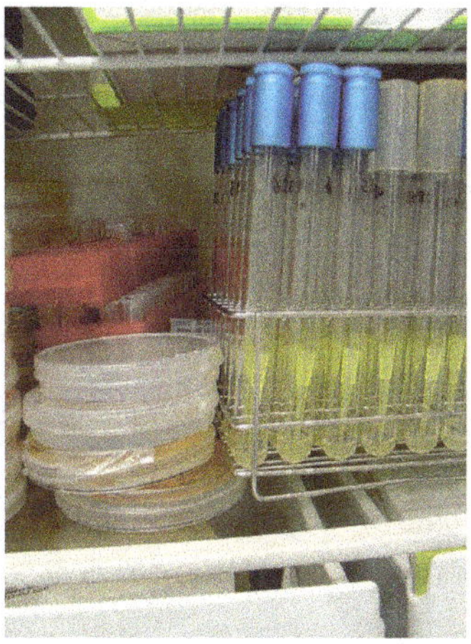

Microbial cultures on solid and liquid media.

It is often essential to isolate a pure culture of microorganisms. A pure (or *axenic*) culture is a population of cells or multicellular organisms growing in the absence of other species or types. A pure culture may originate from a single cell or single organism, in which case the cells are genetic clones of one another. For the purpose of gelling the microbial culture, the medium of agarose gel (agar) is used. Agar is a gelatinous substance derived from seaweed. A cheap substitute for agar is guar gum, which can be used for the isolation and maintenance of thermophiles.

Bacterial Culture

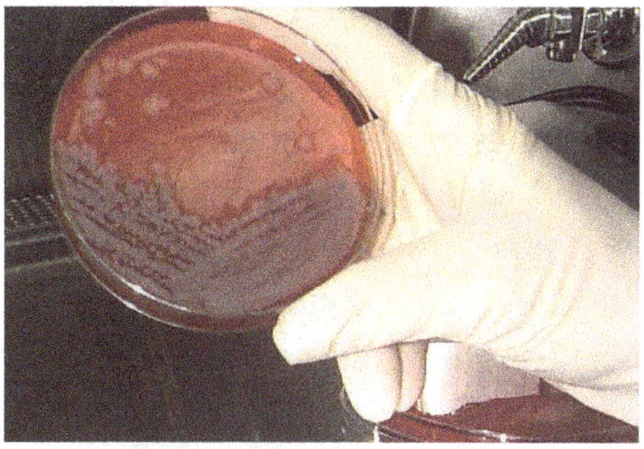

A culture of Bacillus anthracis.

There are several types of bacterial culture methods that are selected based on the agent being cultured and the downstream use.

Broth Cultures

One method of bacterial culture is liquid culture, in which the desired bacteria are suspended in a liquid nutrient medium, such as Luria Broth, in an upright flask. This allows a scientist to grow up large amounts of bacteria for a variety of downstream applications.

Liquid cultures are ideal for preparation of an antimicrobial assay in which the experimenter inoculates liquid broth with bacteria and lets it grow overnight (they may use a shaker for uniform growth). Then they would take aliquots of the sample to test for the antimicrobial activity of a specific drug or protein (antimicrobial peptides).

Liquid cultures of the cyanobacterium *Synechococcus* PCC 7002.

As an alternative, the microbiologist may decide to use static liquid cultures. These cultures are not shaken and they provide the microbes with an oxygen gradient.

Agar Plates

Microbiological cultures can be grown in petri dishes of differing sizes that have a thin layer of agar-based growth medium. Once the growth medium in the petri dish is inoculated with the desired bacteria, the plates are incubated at the optimal temperature for the growing of the selected bacteria (for example, usually at 37 degrees Celsius, or the human body temperature, for cultures from humans or animals, or lower for environmental cultures). After the desired level of growth is achieved, agar plates can be stored upside down in a refrigerator for an extended period of time to keep bacteria for future experiments.

There are a variety of additives that can be added to agar before it is poured into a plate and allowed to solidify. Some types of bacteria can only grow in the presence of certain additives. This can also be used when creating engineered strains of bacteria that contain an antibiotic-resistance gene. When the selected antibiotic is added to the agar, only bacterial cells containing the gene insert conferring resistance will be able to grow. This allows the researcher to select only the colonies that were successfully transformed.

Stab Cultures

Stab cultures are similar to agar plates, but are formed by solid agar in a test tube. Bacteria is introduced via an inoculation needle or a pipette tip being stabbed into the center of the agar. Bacteria grow in the punctured area. Stab cultures are most commonly used for short-term storage or shipment of cultures.

Motile and non-motile bacteria can be differentiated along the stab lines. Motile bacteria will grow out from the stab line while non-motile bacteria are present only along the stab line.

Culture Collections

Microbial culture collections focus on the acquisition, authentication, production, preservation, catalogueing and distribution of viable cultures of standard reference microorganisms, cell lines and other materials for research in microbial systematics. Culture collection are also repositories of type strains.

Major national culture collections.		
Collection Acronym	Name	Location
ATCC	American Type Culture Collection	Manassas, Virginia
BCCM	Belgian Co-ordinated Collections of Micro-organisms	Decentralized, Coordination Cell in Brussels, Belgium
CCUG	Culture Collection University of Gothenburg	Gothenburg, Sweden
CECT	Colección Española de Cultivos Tipo	Valencia, Spain
CIP	Collection d'Institut Pasteur	Paris, France
DSMZ	Deutsche Sammlung von Mikroorganismen und Zellkulturen	Braunschweig, Germany
ICMP	International Collection of Microorganisms from Plants	Auckland, New Zealand
JCM	Japan Collection of Microorganisms	Tsukuba, Ibaraki, Japan
NCTC	National Collection of Type Cultures	Public Health England, London, United Kingdom

Virus and Phage Culture

Virus or phage cultures require host cells in which the virus or phage multiply. For bacteriophages, cultures are grown by infecting bacterial cells. The phage can then be isolated from the resulting plaques in a lawn of bacteria on a plate. Virus cultures are obtained from their appropriate eukaryotic host cells.

Eukaryotic Cell Culture

Isolation of Pure Cultures

For single-celled eukaryotes, such as yeast, the isolation of pure cultures uses the same techniques as for bacterial cultures. Pure cultures of multicellular organisms are often more easily isolated by simply picking out a single individual to initiate a culture. This is a useful technique for pure culture of fungi, multicellular algae, and small metazoa, for example.

Developing pure culture techniques is crucial to the observation of the specimen in question. The most common method to isolate individual cells and produce a pure culture is to prepare a streak plate. The streak plate method is a way to physically separate the microbial population, and is done by spreading the inoculate back and forth with an inoculating loop over the solid agar plate. Upon incubation, colonies will arise and single cells will have been isolated from the biomass.

Types of Culture Media

A culture media is a special medium used in microbiological laboratories to grow different kinds of microorganisms. A growth or a culture medium is composed of different nutrients that are essential for microbial growth.

Since there are many types of microorganisms, each having unique properties and requiring specific nutrients for growth, there are many types based on what nutrients they contain and what function they play in the growth of microorganisms.

A culture may be solid or liquid. The solid culture media is composed of a brown jelly like substance known as agar. Different nutrients and chemicals are added to it to allow the growth of different microorganisms. Below given are some types of important culture or growth media used in microbiological laboratories:

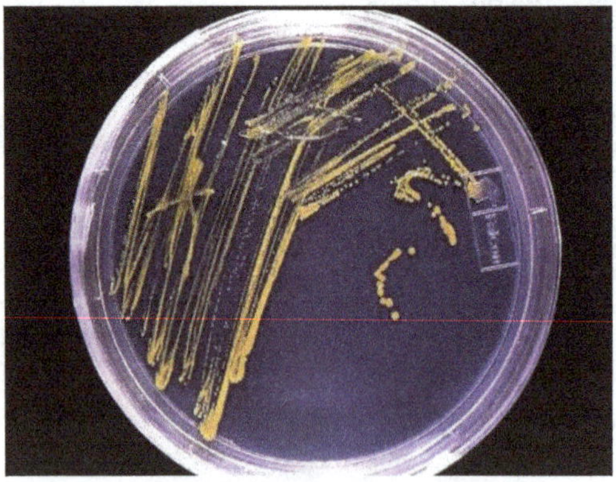

Agar Plate with bacterial colonies.

The Preservation Culture Media

This is composed of all the basic nutrients required for a microbial growth and is used to preserve a specific type of microorganism, preferably bacteria or a set of different microbial entities for a long period of time.

The basic purpose of this culture is to let these microorganisms grow safely in an ensured environment that has all the important nutrients and to protect them against any environmental damage so these organisms can be used when needed.

The Enrichment Culture Media

This is a liquid medium which allows the microorganisms to multiply and has the essential nutrients that are required for it.

It is usually composed of bacteria taken from a liquid source such as pond water. The basic nutrient broth is the most commonly used.

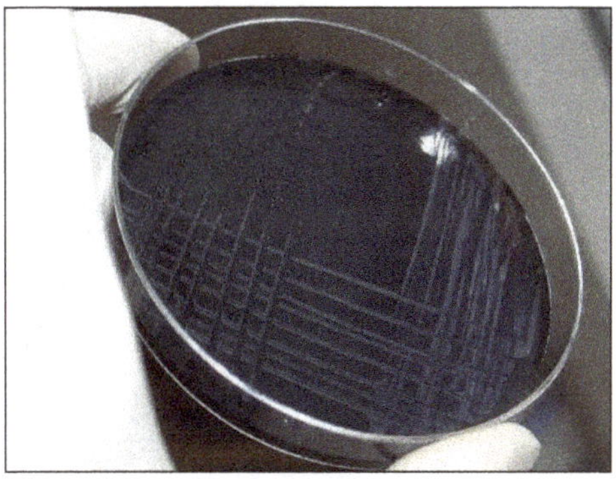

Selective Media Plate.

Selective Culture Media

This is a special type of media which allows the growth of certain microorganisms while inhibits the growth of the others.

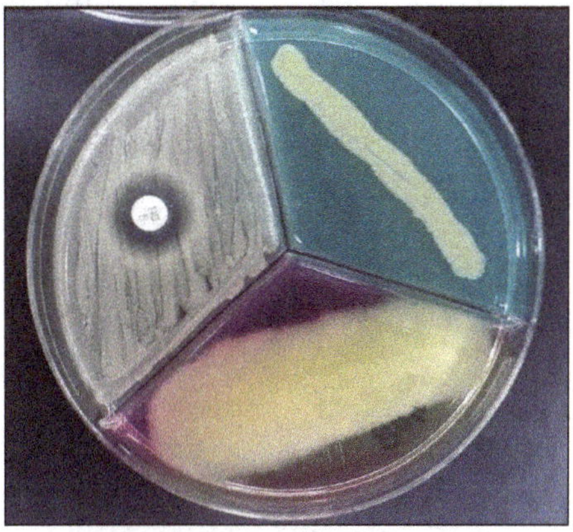

For example if we want to isolate a specific bacteria let's say that can with stand an acidic environment from a sample of pond water and get rid of others, a selective media with a low pH will be taken which will allow the growth of only those organisms that can withstand acidity and will kill the others that cannot.

Examples of commonly used selective media includes: PALCAM agar medium or Mac conkey agar medium.

Differential Culture Media

This is a media that is used for differentiating between bacteria by using an identification marker for a specific type of microorganism.

The selective and differential culture media are opposites to each other in a way that one inhibits the growth of other organisms while allowing the growth of some while the other does not kill the others but only highlights one type.

Blood agar is a common differential culture medium used to identify bacteria that causes haemolysis in blood.

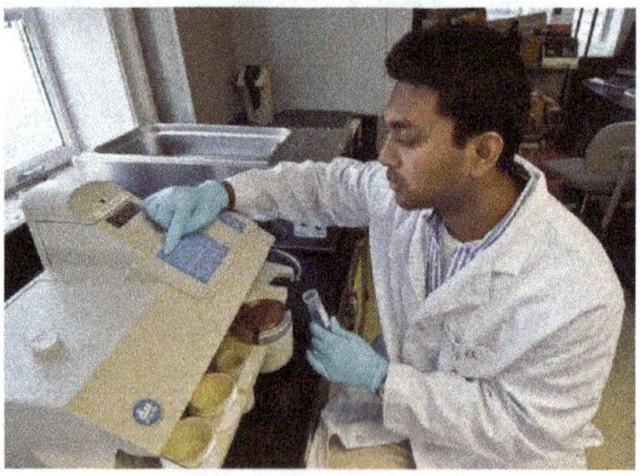

Resuscitation Culture Media

This is a special type of media which is used for growing microorganisms that are damaged and have lost the ability to produce due to certain harmful environmental factors.

Fermentation Media.

This culture allows the organisms to regain their metabolism by providing the nutrients that the organisms had been deprived of. For example, a type of bacteria that requires histamine for its growth is subjected to a medium lacking this essential component its growth will be inhibited.

If the same bacteria is then placed in a medium consisting of histamine it will start to grow again. In this case the media containing histamine will act as resuscitation media. An example of a commonly used resuscitation culture media is the tryptic soya agar.

General Purpose Media

The general purpose media is a media that has a multiple effect, i.e. it can be used as a selective, deferential or a resuscitation media.

Isolation Culture Media

An isolation culture medium is a simple agar containing solid medium that allows the growth of microorganisms in the direction of the streaks.

For example the bacteria will only grow on the pattern made on the solidified agar during the streak plate method. This is the most commonly used medium in microbiological labs.

Fermentation Media

The fermentation culture media is a liquid selective media which is used to obtain a culture of a specific organism more likely yeast or a particular toxin.

The fermentation media can also be differential but mostly it is selective in nature that is allowing the growth of one type while inhibiting the growth of others.

MICROBIAL METABOLISM

Microbial metabolism is the means by which a microbe obtains the energy and nutrients (e.g. carbon) it needs to live and reproduce. Microbes use many different types of metabolic strategies and species can often be differentiated from each other based on metabolic characteristics.

The specific metabolic properties of a microbe are the major factors in determining that microbe's ecological niche, and often allow for that microbe to be useful in industrial processes or responsible for biogeochemical cycles.

Types

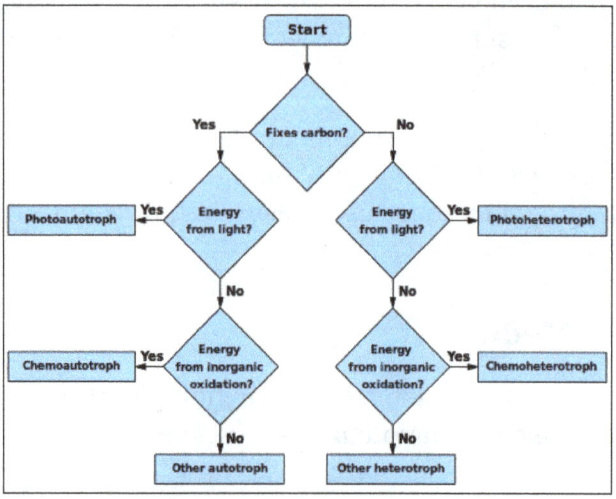

Flow chart to determine the metabolic characteristics of microorganisms.

All microbial metabolisms can be arranged according to three principles:

How the organism obtains carbon for synthesising cell mass:

- Autotrophic: Carbon is obtained from carbon dioxide (CO_2).
- Heterotrophic: Carbon is obtained from organic compounds.
- Mixotrophic: Carbon is obtained from both organic compounds and by fixing carbon dioxide.

How the organism obtains reducing equivalents used either in energy conservation or in biosynthetic reactions:

- Lithotrophic: Reducing equivalents are obtained from inorganic compounds.
- Organotrophic: Reducing equivalents are obtained from organic compounds.

How the organism obtains energy for living and growing:

- Chemotrophic: Energy is obtained from external chemical compounds.
- Phototrophic: Energy is obtained from light.

In practice, these terms are almost freely combined. Typical examples are as follows:

- Chemolithoautotrophs obtain energy from the oxidation of inorganic compounds and carbon from the fixation of carbon dioxide. Examples: Nitrifying bacteria, Sulfur-oxidizing bacteria, Iron-oxidizing bacteria, Knallgas-bacteria.

- Photolithoautotrophs obtain energy from light and carbon from the fixation of carbon dioxide, using reducing equivalents from inorganic compounds. Examples: Cyanobacteria (water (H_2O) as reducing equivalent donor), Chlorobiaceae, Chromatiaceae (hydrogen sulfide (H_2S) as reducing equivalent donor), *Chloroflexus* (hydrogen (H_2) as reducing equivalent donor).

- Chemolithoheterotrophs obtain energy from the oxidation of inorganic compounds, but cannot fix carbon dioxide (CO_2). Examples: some *Thiobacilus*, some *Beggiatoa*, some *Nitrobacter* spp., *Wolinella* (with H_2 as reducing equivalent donor), some Knallgas-bacteria, some sulfate-reducing bacteria.

- Chemoorganoheterotrophs obtain energy, carbon, and reducing equivalents for biosynthetic reactions from organic compounds. Examples: most bacteria, e. g. *Escherichia coli*, *Bacillus* spp., *Actinobacteria*.

- Photoorganoheterotrophs obtain energy from light, carbon and reducing equivalents for biosynthetic reactions from organic compounds. Some species are strictly heterotrophic, many others can also fix carbon dioxide and are mixotrophic. Examples: *Rhodobacter*, *Rhodopseudomonas*, *Rhodospirillum*, *Rhodomicrobium*, *Rhodocyclus*, *Heliobacterium*, *Chloroflexus* (alternatively to photolithoautotrophy with hydrogen).

Heterotrophic Microbial Metabolism

Some microbes are heterotrophic (more precisely chemoorganoheterotrophic), using organic compounds as both carbon and energy sources. Heterotrophic microbes live off of nutrients that they scavenge from living hosts (as commensals or parasites) or find in dead organic matter of all kind (saprophages). Microbial metabolism is the main contribution for the bodily decay of all organisms after death. Many eukaryotic microorganisms are heterotrophic by predation or parasitism, properties also found in some bacteria such as *Bdellovibrio* (an intracellular parasite of other bacteria, causing death of its victims) and Myxobacteria such as *Myxococcus* (predators of other bacteria which are killed and lysed by cooperating swarms of many single cells of Myxobacteria). Most pathogenic bacteria can be viewed as heterotrophic parasites of humans or the other eukaryotic species they affect. Heterotrophic microbes are extremely abundant in nature and are responsible for the breakdown of large organic polymers such as cellulose, chitin or lignin which are generally indigestible to larger animals. Generally, the breakdown of large polymers to carbon dioxide (mineralization) requires several different organisms, with one breaking down the polymer into its constituent monomers, one able to use the monomers and excreting simpler waste compounds as by-products, and one able to use the excreted wastes. There are many variations on this theme, as different organisms are able to degrade different polymers and secrete different waste products. Some organisms are even able to degrade more recalcitrant compounds such as petroleum compounds or pesticides, making them useful in bioremediation.

Biochemically, prokaryotic heterotrophic metabolism is much more versatile than that of eukaryotic organisms, although many prokaryotes share the most basic metabolic models with eukaryotes, e. g. using glycolysis (also called EMP pathway) for sugar metabolism and the citric acid cycle to degrade acetate, producing energy in the form of ATP and reducing power in the form of NADH or quinols. These basic pathways are well conserved because they are also involved in biosynthesis of many conserved building blocks needed for cell growth (sometimes in reverse direction). However, many bacteria and archaea utilize alternative metabolic pathways other than glycolysis and the citric acid cycle. A well-studied example is sugar metabolism via the keto-deoxy-phosphogluconate pathway (also called ED pathway) in *Pseudomonas*. Moreover, there is a third alternative sugar-catabolic pathway used by some bacteria, the pentose phosphate pathway. The metabolic diversity and ability of prokaryotes to use a large variety of organic compounds arises from the much deeper evolutionary history and diversity of prokaryotes, as compared to eukaryotes. It is also noteworthy that the mitochondrion, the small membrane-bound intracellular organelle that is the site of eukaryotic energy metabolism, arose from the endosymbiosis of a bacterium related to obligate intracellular *Rickettsia*, and also to plant-associated *Rhizobium* or *Agrobacterium*. Therefore, it is not surprising that all mitrochondriate eukaryotes share metabolic properties with these Proteobacteria. Most microbes respire (use an electron transport chain), although oxygen is not the only terminal electron acceptor that may be used. As discussed below, the use of terminal electron acceptors other than oxygen has important biogeochemical consequences.

References

- Madigan, Michael T. (2012). Brock biology of microorganisms (13th ed.). San Francisco: Benjamin Cummings. ISBN 9780321649638
- Morphology-of-microbial-cell-microbiology, cell: biologydiscussion.com, Retrieved 17 April, 2020
- Uruburu, F. (2003). "History and services of culture collections". International Microbiology. 6 (2): 101–103. Doi:10.1007/s10123-003-0115-2. PMID 12811589
- Sterilization, micronotes: microrao.com, Retrieved 18 May, 2020
- Schink B, Thiemann V, Laue H, Friedrich MW (May 2002). "Desulfotignum phosphitoxidans sp. Nov., a new marine sulfate reducer that oxidizes phosphite to phosphate". Arch Microbiol. 177 (5): 381–91. Doi:10.1007/s00203-002-0402-x. PMID 11976747
- Types-Of-Culture-Media, stem: owlcation.com, Retrieved 19 June, 2020
- Engel, A. Et. Al (2011). "Effects of sea surface warming on the production and composition of dissolved organic matter during phytoplankton blooms: results from a mesocosm study." Journal of Plankton Research. Vol 33 Issue 3, p357-372, 160p
- West SA, Buckling A. 2003. Cooperation, virulence and siderophore production in bacterial parasites. Proc. R. Soc. Lon. Ser. B 270:37–44. Doi: 10.1098/rspb.2002.2209
- Fenchel T. (2008). "The microbial loop – 25 years later". Journal of Experimental Marine Biology and Ecology. 99-103

Important Theories in Microbiology

5 CHAPTER

Theory of abiogenesis, theory of biogenesis, microbial theory of fermentation, hologenome concept of evolution and germ theory of disease are some of the theories that fall under the domain of microbiology. This chapter has been carefully written to provide an easy understanding of these theories of microbiology.

THEORY OF ABIOGENESIS

Abiogenesis is the idea that life arose from nonlife more than 3.5 billion years ago on Earth. Abiogenesis proposes that the first life-forms generated were very simple and through a gradual process became increasingly complex. Biogenesis, in which life is derived from the reproduction of other life, was presumably preceded by abiogenesis, which became impossible once Earth's atmosphere assumed its present composition.

Although many equate abiogenesis with the archaic theory of spontaneous generation, the two ideas are quite different. According to the latter, complex life (e.g., a maggot or mouse) was thought to arise spontaneously and continually from nonliving matter. While the hypothetical process of spontaneous generation was disproved as early as the 17th century and decisively rejected in the 19th century, abiogenesis has been neither proved nor disproved.

The Oparin-haldane Theory

In the 1920s British scientist J.B.S. Haldane and Russian biochemist Aleksandr Oparin independently set forth similar ideas concerning the conditions required for the origin of life on Earth. Both believed that organic molecules could be formed from abiogenic materials in the presence of an external energy source (e.g., ultraviolet radiation) and that the primitive atmosphere was reducing (having very low amounts of free oxygen) and contained ammonia and water vapour, among other gases. Both also suspected that the first life-forms appeared in the warm, primitive ocean and were heterotrophic (obtaining preformed nutrients from the compounds in existence on early Earth) rather than autotrophic (generating food and nutrients from sunlight or inorganic materials).

Oparin believed that life developed from coacervates, microscopic spontaneously formed spherical aggregates of lipid molecules that are held together by electrostatic forces and that may have been precursors of cells. Oparin's work with coacervates confirmed that enzymes fundamental for the biochemical reactions of metabolism functioned more efficiently when contained within membrane-bound spheres than when free in aqueous solutions. Haldane, unfamiliar with Oparin's coacervates, believed that simple organic molecules formed first and in the presence of ultraviolet light became increasingly complex, ultimately forming cells. Haldane and Oparin's ideas formed the foundation for much of the research on abiogenesis that took place in later decades.

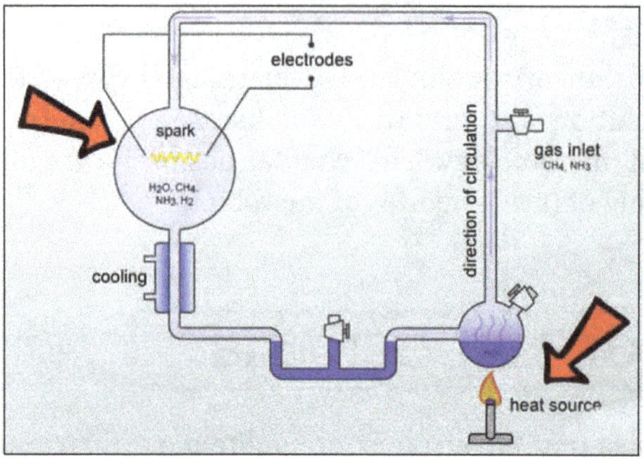

The Miller-urey Experiment

In 1953 American chemists Harold C. Urey and Stanley Miller tested the Oparin-Haldane theory and successfully produced organic molecules from some of the inorganic components thought to have been present on prebiotic Earth. In what became known as the Miller-Urey experiment, the two scientists combined warm water with a mixture of four gases—water vapour, methane, ammonia, and molecular hydrogen—and pulsed the "atmosphere" with electrical discharges. The different components were meant to simulate the primitive ocean, the prebiotic atmosphere, and heat (in the form of lightning), respectively. One week later Miller and Urey found that simple organic molecules, including amino acids (the building blocks of proteins), had formed under the simulated conditions of early Earth.

Modern Conceptions of Abiogenesis

Modern abiogenesis hypotheses are based largely on the same principles as the Oparin-Haldane theory and the Miller-Urey experiment. There are, however, subtle differences between the several models that have been set forth to explain the progression from abiogenic molecule to living organism, and explanations differ as to whether complex organic molecules first became self-replicating entities lacking metabolic functions or first became metabolizing protocells that then developed the ability to self-replicate.

The habitat for abiogenesis has also been debated. While some evidence suggests that life may have originated from nonlife in hydrothermal vents on the ocean floor, it is possible that abiogenesis occurred elsewhere, such as deep below Earth's surface, where newly arisen protocells could have subsisted on methane or hydrogen, or even on ocean shores, where proteinoids may have emerged from the reaction of amino acids with heat and then entered the water as cell-like protein droplets.

Some scientists have proposed that abiogenesis occurred more than once. In one example of this hypothetical scenario, different types of life arose, each with distinct biochemical architectures reflecting the nature of the abiogenic materials from which they developed. Ultimately, however, phosphate-based life ("standard" life, having a biochemical architecture requiring phosphorus) gained an evolutionary advantage over all non-phosphate-based life ("nonstandard" life) and thereby became the most widely distributed type of life on Earth. This notion led scientists to infer the existence of a shadow biosphere, a life-supporting system consisting of microorganisms of unique or unusual biochemical structure that may have once existed, or possibly still exists, on Earth.

As the Miller-Urey experiment demonstrated, organic molecules can form from abiogenic materials under the constraints of Earth's prebiotic atmosphere. Since the 1950s, researchers have found that amino acids can spontaneously form peptides (small proteins) and that key intermediates in the synthesis of RNA nucleotides (nitrogen-containing compounds [bases] linked to sugar and phosphate groups) can form from prebiotic starting materials. The latter evidence may support the RNA world hypothesis, the idea that on early Earth there existed an abundance of RNA life produced through prebiotic chemical reactions. In fact, in addition to carrying and translating genetic information, RNA is a catalyst, a molecule that increases the rate of a reaction without itself being consumed, meaning that a single RNA catalyst could have produced multiple living forms, which would have been advantageous during the rise of life on Earth. The RNA world hypothesis is one of the leading self-replication-first conceptions of abiogenesis.

Blue-green algae in Morning Glory Pool, Yellowstone National Park, Wyoming.

Some modern metabolism-based models of abiogenesis incorporate Oparin's enzyme-containing coacervates but suggest a steady progression from simple organic molecules to coacervates, specifically protobionts, aggregates of organic molecules that display some characteristics of life. Protobionts presumably then gave rise to prokaryotes, single-celled organisms lacking a distinct nucleus and other organelles because of the absence of internal membranes but capable of metabolism and self-replication and susceptible to natural selection. Examples of primitive prokaryotes still found on Earth today include archaea, which often inhabit extreme environments with conditions similar to those that may have existed billions of years ago, and cyanobacteria (blue-green algae), which also flourish in inhospitable environments and are of particular interest in understanding the origin of life, given their photosynthetic abilities. Stromatolites, deposits formed by the growth of blue-green algae, are the world's oldest fossils, dating to 3.5 billion years ago.

There remain many unanswered questions concerning abiogenesis. Experiments have yet to demonstrate the complete transition of inorganic materials to structures like protobionts and protocells and, in the case of the proposed RNA world, have yet to reconcile important differences in mechanisms in the synthesis of purine and pyrimidine bases necessary to form complete RNA nucleotides. In addition, some scientists contend that abiogenesis was unnecessary, suggesting instead that life was introduced on Earth via collision with an extraterrestrial object harbouring living organisms, such as a meteorite carrying single-celled organisms; the hypothetical migration of life to Earth is known as panspermia.

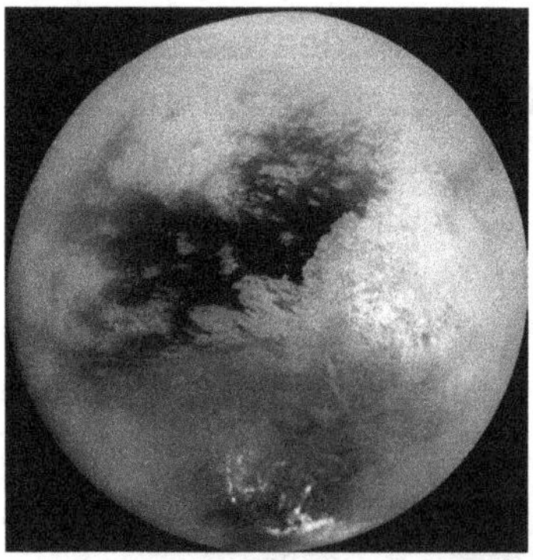

Saturn's moon Titan, in a mosaic of nine images taken by the Cassini spacecraft on October 26, 2004, and processed to reduce the veiling effects of the moon's atmosphere. The view is centred slightly south of the equator, with north toward the top. The continent-size region Xanadu Regio shows as the large bright patch on the right, while bright methane clouds appear near Titan's south pole.

THEORY OF BIOGENESIS

The theory of biogenesis states that living things can only arise from living things and cannot be spontaneously generated.

The Spontaneous generation hypothesis proposed by scientists to explain the origin of the "animalcules" observed by Antoni van Leeuwenhoek in his magnifying lenses had received wide acceptance all over Europe from Antoni's time until the time of Louis Pasteur. Erroneous experimental set up, results, and conclusions of some scientists had supported and strengthened the hypothesis.

For example, the Englishman John Needham claimed that vital life is needed for the spontaneous generation of microbes. He added that the reason why no living organisms emerged from heated and sealed solutions in containers is that the "vital life" was destroyed by the heat and new "vital life" was not supplied to the solutions because they cannot enter the sealed containers.

Fortunately, there were scientists skeptical about the hypothesis, so they designed their own experimental set up and from the results they gathered, they drew the most feasible explanation on the origin of the "animalcules". Among the scientists was the Italian Lazzaro Spallanzani who opposed Needham's idea of the "vital life".

Proponents and opponents of spontaneous generation hypothesis debated a lot starting from the time Leeuwenhoek presented his discoveries to the public until the time of Rudolf Virchow, who in 1858 challenged the spontaneous generation with his concept and definition of biogenesis.

This concept claims that living cells can arise only from preexisting living cells. Virchow defended this concept to the scientific community but he did not come up with a convincing experiment to back up his idea. In 1861, the French scientist Louis Pasteur resolved the issue on the origin of microbes ("animalcules") through a series of ingenious and persuasive experiments.

Pasteur showed that microorganisms exist in the air and can contaminate sterile solutions, but he emphasized that air itself does not produce microbes. He filled a number of short-necked flasks with beef broth and then boiled their contents. He immediately sealed the mouths of some of the flasks while he left the others open and allowed to cool.

After few days, the contents of the unsealed flasks were found to be contaminated with microorganisms. No evidences of growing microorganisms were found on the sealed flasks. Pasteur concluded that the microorganisms in the air were responsible for contaminating non-living matter like the broths in John Needham's flask.

Pasteur performed another experiment but this time he put beef broth in open-ended long-necked flasks. He bent the necks of the flasks into S-shaped curves and boiled the contents of the flasks. Amazingly, the contents of the flasks were not contaminated even after several months.

The unique S-shaped design of Pasteur's flasks allowed air to pass but trap microorganisms that may contaminate the broths. Do you know that some of the original vessels used by Pasteur in his experiments are still displayed in the Pasteur Institute, Paris today? A few of the flasks contain broths that remain uncontaminated for more than 100 years.

Pasteur demonstrated the presence of microbes in non-living materials whether they are solid, liquid, or air. In addition, he laid the foundation of aseptic techniques, techniques that prevent contamination by unwanted microbes.

These techniques are based on Pasteur's idea that microbes can be killed by heat and that procedures can be designed to inhibit the access of airborne microbes to nutrient environment. Application of aseptic techniques is now the standard practice in medical and laboratory procedures.

Disproving the idea that microorganisms spontaneously generated from non-living matter through mystical forces is one of the greatest contributions of Pasteur in science. He provided the evidence that any appearance of "spontaneous" life in nonliving solutions can be attributed to microbes that already exist in the air or in the fluids themselves.

Belief in Spontaneous Generation

Spontaneous generation is also known as Aristotelian abiogenesis, after its ancient Greek proponent. The stealth and invisibility of such organisms as flies, mice and bacteria allowed belief in spontaneous generation to hold sway for millennia. Pioneering use of the still-new microscope in the 18th century began to erode its credibility; seeing fly eggs and bacteria under the microscope helped to demystify their nature. By Pasteur's time, experimentation had defended biogenesis at the macroscopic level. Only microscopic biogenesis was left to be proven.

Macroscopic Spontaneous Generation

In 1668, Francesco Redi addressed the question of macroscopic spontaneous generation when he published the results of an experiment in which he placed rotting meat in a container and covered the container's opening with gauze. If the gauze was absent, maggots would grow on the meat. If the gauze was present, maggots would not grow on the meat, but would appear on the gauze. Redi was observing flies depositing eggs as close to a food source as could be reached.

Microscopic Spontaneous Generation

A century later, an experiment performed by Lazzaro Spallanzani in 1768 indicated biogenesis on the microscopic level. Spallanzani wanted to avoid contamination by boiling a meat broth in a sealed container. The problem with this approach was that air in the container could shatter the container upon heating. Therefore, he evacuated the container after sealing it closed. The broth did not subsequently cloud with bacterial growth, supporting the theory of biogenesis.

Critics charged that air is needed for life. The lack of bacterial growth therefore was presumed due to a lack of air, not because bacteria spread through contamination.

This criticism stood for almost a century before Pasteur entered the scene and overturned it.

Pasteur's Experimental Equipment

The 1859 experiment performed by Pasteur unequivocally overturned the theory of spontaneous generation at the microscopic level. He boiled a meat broth in a flask that had a long neck that curved downward, then upward, like a goose neck. The bend in the neck prevented contaminating particles from reaching the broth, while still allowing the free diffusion of air. The fact that the flask allowed for the passage of air was a design breakthrough that finally addressed the critics of Spallanzani.

Pasteur's flask remained free of bacterial growth for as long as the flask remained upright. To show where the contaminating elements were located, he tipped the flask enough for the broth to sweep out the bend in the goose neck; the broth would then quickly become clouded with bacterial growth.

MICROBIAL THEORY OF FERMENTATION

The fermentation theory was studied in depth and brought to light first by Louis Pasteur. This theory states that it is the idea or concept of how fermentation is brought on by microbes and put to the concept of spontaneous generation to rest. Even though this theory is now outdated and has been replaced by the germ theory of disease, for a long time it held true, and Louis was on the forefront of explaining why it seemed organisms appeared out of nothing instead of claiming it was just a spontaneous act of God. Fermentation was a process that has been used for thousands of years, but no one could explain exactly what was happening and why. From Pasteur's discovery of why and how fermentation occurs, the process has been studied intensely and is now a mastered art used in everyday life with processes of making things such as alcoholic beverages, some foods like yogurt or even manufacturing some medications.

Louis Pasteur, Archives Photographiques.

Fermentation

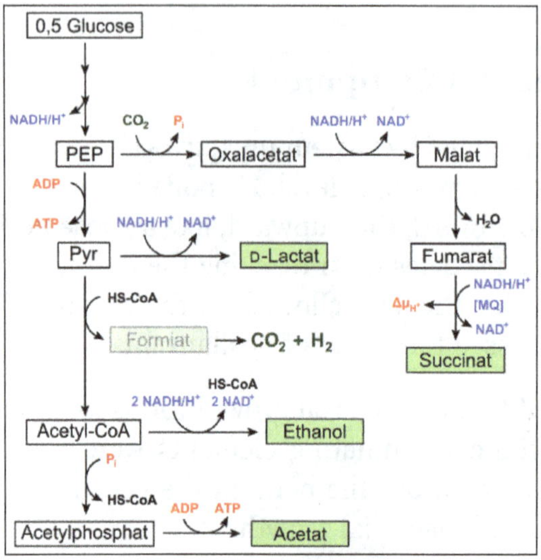

Process of Fermentation.

Simply put, fermentation is the anaerobic metabolic process that converts sugar into acids, gases, or alcohols. This metabolic process is used in oxygen starved environments. Yeast and many other microbes commonly use this process in order to carry our their anaerobic respiration to survive. Even the human body carries out fermentation processes from time to time. When someone runs a long distance race, lactic acid will build up in their muscles over the course of the race. That lactic acid is the by-product of fermentation taking place in their body, which tries to produce ATP so the body can continue to run since they could not process the oxygen intake fast enough. Although fermentation will give a lower yield of ATP production than aerobic respiration does, it can occur at a much higher rate. Fermentation has been used by humans consciously since around 5000 BCE where there were jars recovered in the Iran Zagros Mountains area in which contained remnants of a microbes similar those present in the process of making grapes into wine.

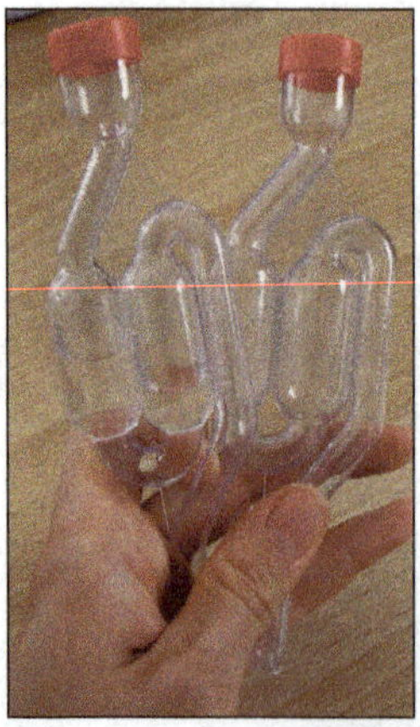

Example of a typical curved neck used today.

Theory

Before the 1870s, when Pasteur published his work on this theory, it was believed that microorganisms and even some small animals such as frogs would spontaneously appear, which was coined as spontaneous generation. Spontaneous generation was the explained theory that when elements of the Earth such as clay or mud would mix with water and sunlight in certain amounts, creatures would just appear out of that concoction. A common way that this idea was "proven" over and over again was by taking a piece of raw meat and placing it in open air, which would almost always produce maggots. This idea was accepted and believe to be true before Louis Pasteur shook the Earth with his new ideas that organisms actually came from traceable beginnings. Pasteur demonstrated that fermentation is caused by the growth of microorganisms, and the emergent growth of bacteria in nutrient broths is due to biogenesis rather than spontaneous generation. He exposed boiled broths to air in vessels that contained a filter to prevent all particles from passing through to the growth medium. Yet, when the vessels were open to the air surrounding it, the organisms appeared. It could be concluded that spontaneous generation could be disproven. The organisms did not just appear but were coming from the air, yet we were not able to see them at such a small level. His famous experiment was used with a curved neck placed on top of a beaker. This curved neck was the key to proving his findings because it showed that the germs and microbes had to fall into the broth inside. The curved neck did not allow this to happen.

The fermentation theory of disease is the (now obsolete) concept that many diseases, including the diseases which were "epidemic, endemic and contagious", owe their origin to the presence of a "morbific principle" in the system, acting in a manner analogous to, although not identical with, the process of fermentation. It was rendered obsolete by the germ theory of disease, which led to the new science of bacteriology.

Application

Fermenting a broth in a beaker has become more than just a way to prove that organisms don't just appear out of thin air. Today, the process of fermentation is used for a multitude of everyday applications. Some of those processes include medications which we ingest, beverages we consume, and even food we eat. Currently, companies like Genencor International uses the production of enzymes involved in fermentation to build a revenue of over $400 million a year. Many medications such as antibiotics are produced by the fermentation process. An example is the important drug cortisone, which can be prepared by the fermentation of a plant steroid known as diosgenin.

The enzymes used in the reaction are provided by the mold Rhizopus nigricans. Just as it is commonly known, alcohol of all types and brands are also produced by way of fermentation and distillation. Moonshine is a classic example of how this is carried out.

Finally, foods such as yogurt are made by fermentation processes as well. Yogurt is a fermented milk product that contains the characteristic bacterial cultures Lactobacillus bulgaricus and Streptococcus thermopiles.

Burton Union fermentation system, Coors Visitor Centre.

References

- Abiogenesis, science: britannica.com, Retrieved 20 July, 2020
- Pasteur, Louis. "Physiological Theory of Fermentation". Fordham University. Retrieved March 13, 2020
- Biogenesis-theory-in-microbiology, molecular-biology, topics: bioscience.com.pk, Retrieved 21 August, 2020
- Theory-biogenesis: sciencing.com, Retrieved 22 January, 2020
- Germ-theory: biologydictionary.net, Retrieved 23 February, 2020

PERMISSIONS

All chapters in this book are published with permission under the Creative Commons Attribution Share Alike License or equivalent. Every chapter published in this book has been scrutinized by our experts. Their significance has been extensively debated. The topics covered herein carry significant information for a comprehensive understanding. They may even be implemented as practical applications or may be referred to as a beginning point for further studies.

We would like to thank the editorial team for lending their expertise to make the book truly unique. They have played a crucial role in the development of this book. Without their invaluable contributions this book wouldn't have been possible. They have made vital efforts to compile up to date information on the varied aspects of this subject to make this book a valuable addition to the collection of many professionals and students.

This book was conceptualized with the vision of imparting up-to-date and integrated information in this field. To ensure the same, a matchless editorial board was set up. Every individual on the board went through rigorous rounds of assessment to prove their worth. After which they invested a large part of their time researching and compiling the most relevant data for our readers.

The editorial board has been involved in producing this book since its inception. They have spent rigorous hours researching and exploring the diverse topics which have resulted in the successful publishing of this book. They have passed on their knowledge of decades through this book. To expedite this challenging task, the publisher supported the team at every step. A small team of assistant editors was also appointed to further simplify the editing procedure and attain best results for the readers.

Apart from the editorial board, the designing team has also invested a significant amount of their time in understanding the subject and creating the most relevant covers. They scrutinized every image to scout for the most suitable representation of the subject and create an appropriate cover for the book.

The publishing team has been an ardent support to the editorial, designing and production team. Their endless efforts to recruit the best for this project, has resulted in the accomplishment of this book. They are a veteran in the field of academics and their pool of knowledge is as vast as their experience in printing. Their expertise and guidance has proved useful at every step. Their uncompromising quality standards have made this book an exceptional effort. Their encouragement from time to time has been an inspiration for everyone.

The publisher and the editorial board hope that this book will prove to be a valuable piece of knowledge for students, practitioners and scholars across the globe.

INDEX

CPSIA information can be obtained
at www.ICGtesting.com
Printed in the USA
BVHW062118290822
645775BV00003B/183

9 781639 893546